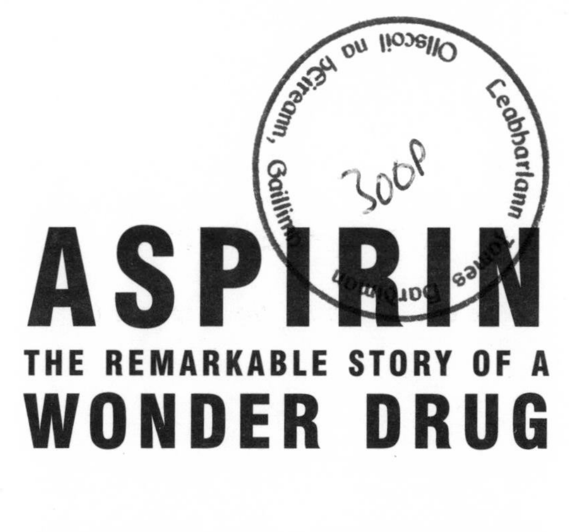

ASPIRIN

THE REMARKABLE STORY OF A WONDER DRUG

DIARMUID JEFFREYS

BLOOMSBURY

First published in Great Britain in 2004

Bloomsbury Publishing Plc, 38 Soho Square, London W1D 3HB

A CIP catalogue record for this book is available from the British Library

Hardback ISBN 0 7475 7077 9
Export Paperback ISBN 0 7475 7443 X

10 9 8 7 6 5 4 3 2 1

Typeset by Palimpsest Book Production Limited, Polmont, Stirlingshire
Printed in Great Britain by Clays Ltd, St Ives plc

All papers used by Bloomsbury Publishing are natural, recyclable products made from wood grown in well-managed forests. The manufacturing processes conform to the environmental regulations of the country of origin

For my parents

CONTENTS

ACKNOWLEDGEMENTS

I COULD NOT have written this book without the help, support and encouragement of a great many people and I am glad to have this opportunity to record my huge debt of gratitude to them.

My thanks, then, to Mark Nesbitt of the Centre for Economic Botany at Kew and John Larsen at the Oriental Institute in Chicago. Both were generous with advice about their specialist areas of knowledge and guided me gently towards sources and information that I would simply not have found on my own. The Reverend Ralph Mann welcomed me into his home and shared his years of scholarly research into the life and times of Edward Stone, giving me many invaluable insights in the process. Sir John Vane was immeasurably helpful. It must have been a long time since he last encountered someone who was quite so scientifically illiterate, but he took my ignorance in his stride and good-naturedly let me plunder his memories about his early life, his career as a scientist and his Nobel Prize-winning discoveries. Peter Elwood was kindness personified, leading me through his story and answering my many questions with great patience and charm. Joseph Collier helped me understand what made Harry Collier tick and why he was so fascinated by aspirin. Ernst Eichengrün told me things about his grandfather that no amount of digging around in the archives would have revealed. Eva Mozes Kor allowed me a glimpse into the horrors of Auschwitz – although I doubt if anyone who wasn't there will ever truly understand how dreadful it must have been.

There were others, too: Gareth Morgan, Charles Dibble, Sir Richard Peto, Chris Paraskeva, Jennifer Tait, Kathryn Uhrich, Chuck Jones, Arslan Akhmedkhanov, Walter Sneader, Phillipe Bellon, Richard Fazzini, Dee Baring, Stephen Nicholas and many more like them, too numerous to mention, who met me, answered my phone calls, replied to my emails, put me right about facts, sent me documents or simply pointed me in the right direction. I am sure there will be dozens of doctors, scientists and experts of one sort or another who will be delighted that I've now stopped badgering them, but this book has been written on the back of their accumulated knowledge and wisdom and I am deeply in their debt.

I am also enormously grateful to the archivists at Bayer AG in Leverkusen, near Cologne, particularly Hans-Herman Pogarell and Rudiger Borstel who gave me access to company records and dealt promptly and courteously with all my subsequent queries. Likewise to Gordon Stephenson at Hull's Reckitt's Heritage centre (now part of Reckitt Benckiser Healthcare) for all his help and for finding George Colman Green's Disprin story for me.

Elsewhere, the British Library, the British Newspaper Library, the US Library of Congress and the wonderful Wellcome Library for the History and Understanding of Medicine far exceeded my expectations, both in terms of the size and scale of their collections of relevant material and the help that their staff gave me to find what I needed. I'm hugely grateful, too, to Mark Nash who tracked down documents in various American archives that I was unable to find myself, to my brother, Mike, who helped me obtain old medical papers here in the UK, and to Karl Hause who did the same in Germany and then had to translate them into English as well.

My heartfelt thanks also to Bill Swainson, my editor at Bloomsbury, for saying yes when the proposal for this book thudded on to his desk, for being so patient and good-humoured throughout and for reminding me gently but firmly, that sometimes 'less is more'. I also want to say a particularly warm thank you to my dear friend and agent, Anthony Sheil. Yet again he proved himself to be a rock.

On a personal note, my friends, my siblings and my parents all gave me their unstinting support, help and encouragement – at times far beyond what I could ever have hoped for or expected – and my thanks go to them all. Finally, to my children Laura and Joe and to my wife Patsy, who walked every faltering step of the way by my side, I give my enduring love and gratitude. Without them it would have been impossible.

PREFACE

SOMEWHERE CLOSE BY – in the bathroom cabinet, tucked away in a desk drawer, at the bottom of an old jacket pocket – there'll be a container with some aspirin in it. Take out a tablet and examine it for a moment.

It's a pretty innocuous-looking thing, isn't it? Just an ordinary little white pill. You've seen hundreds of them before and no doubt you'll see hundreds of them again. It's nothing special.

Look again and consider this. What you're holding is one of the most amazing creations in medical history, a drug so astonishingly versatile that it can relieve your headache, ease your aching limbs, lower your temperature *and* treat some of the deadliest human diseases. There's now evidence to show that aspirin might prevent heart attacks, strokes, deep vein thrombosis, bowel, lung and breast cancer, cataracts, migraine, infertility, herpes, Alzheimer's disease, and much else. The list is growing every year – which might go some way towards explaining why over 25,000 scientific papers have been written about aspirin and why an estimated one trillion little white pills, just like yours, have been consumed since it first came into being.

In short, what you have there is a wonder drug, something with few equivalents in the annals of medical science, and one of the most endurably successful commercial products of all time.

I first became interested in this extraordinary substance several years ago when my father had a serious heart attack. Fortunately, to the great relief of all of us who love him, he recovered and is now fit and well. But since then he's taken a small aspirin tablet daily to keep the blood flowing through his arteries. In recent years, my mother has started taking aspirin too, for much the same reason, and the two of them now follow a strict regimen of one 75 mg dose every day.

It got me wondering. Where did this drug come from? How did

something evolve from an ordinary headache pill into a drug that can save lives? The little I knew about aspirin's origins came from a half-remembered lesson at school – namely, that a German chemical company had invented it sometime in the nineteenth century.

But when I began to look into it I realized that here was a far richer, more complex and much older tale; that this little chemical marvel had the most incredible history. Out of that realization grew this book.

Although we now take it for granted, aspirin didn't just appear from nowhere. It's a product of a roller-coaster ride through time, of accidental discoveries and intuitive reasoning, of astounding scientific ingenuity, personal ambition and intense corporate rivalry. Its past embraces, in no particular order: wars, epidemics, an Oxfordshire vicar, a forgotten Jewish scientist, ancient papyrus scrolls, the Industrial Revolution, a nineteenth-century Scottish fever hospital, a common tree, espionage, a mighty German industrialist, malaria, the Treaty of Versailles, the colour mauve, the City of Hull, Lily the Pink, the world's most powerful pharmaceutical companies, a twitching rabbit aorta, Auschwitz, a mercurial advertising genius, and a great deal more.

I found that there was nothing predetermined about the way all these things came together or the impact they had. If chance had not sometimes played a huge role in its development, aspirin might never have come into existence. Had it not thrived through profit-driven competition, it might never have survived to release its incredible therapeutic benefits. But come together they did – a great mosaic of people, places and events that combined, little by little, to give us one of the most remarkable inventions in history.

This, then, is aspirin's story. And like all really good stories it began a long time ago . . .

BOOK ONE

1

IF YOU EXAMINE A MAN . . .

IT HAD BEEN a long afternoon. They had sat down when the midday sun was flooding the room; now the only light came from an oil lamp swaying gently over their heads. It cast strange flickering shadows on to the walls and over the two heaps of objects that lay on the low table in front of them. These were evidence of their cautious preliminary fencing – the half-dozen scarabs and amulets that the American had been persuaded to buy, and a much larger pile of those he hadn't. But now they had arrived at the critical point in their discussions, the real reason why they were there and the moment for which, like all good salesmen, they had bided their time. As one of them untied the linen wrapper from around the scrolls, his brother watched the American's face for any flicker of interest. They knew he had a weakness for these pieces of old papyrus and had sold him some before, but he could drive a hard bargain and was quite capable of rejecting anything he considered worthless or uninteresting. Unfortunately, the American was the only one among them who could even attempt to decipher the pictures and symbols that covered the papyri, and so they had to wait and gauge his reaction before they knew if they could expect a deal.

For his part, Edwin Smith struggled to conceal his excitement. His landlord Mustapha Agha, watching impassively from a corner, had warned him to expect something special. Smith knew, of course, that Agha would be getting a healthy commission on anything sold that evening and that it was therefore in his interest to inflate the value of the artefacts on offer. But the brothers Ahmed and Mohammed Abd er-Rasul were different from the ordinary touts who pestered rich tourists in Luxor market. They were the city's most successful grave-robbers. For years they'd been Agha's main suppliers of *antikas* and they invariably came up with something worthwhile. Even their fakes – which they knocked up between forays underground – were remarkably convincing. Smith had bought and sold on enough of them himself to know that.

In any case, if what he had been told earlier was true, these papyri really could be unique. Agha had said they had been found between the feet of a mummy in the Assassif area of the Theban necropolis, on the other side of the Nile. Smith knew that the brothers had found their way into a number of officially 'undiscovered' tombs at Assassif – he had obtained some of his best artefacts from that source. If these scrolls were genuinely from the same place, they would bear close examination.

He leant forward and peered intently at the hieratic text, revealed as the first scroll was opened out under the lamplight. It was long, one of the longest he'd ever seen, and covered with a bewildering array of symbols. But they were written in a clear hand; the nameless author must have been a practised and painstaking scribe. Smith murmured to himself, lost in the beauty of a document that had survived almost three millennia. He began to pick out the meaning of some of the words . . . 'If you examine a man . . .'

The three Egyptians watched him, exchanged meaningful glances and then settled back comfortably to wait. They knew now that their patience would be rewarded.

Edwin Smith was born in Bridgeport, Connecticut in April 1822. Although details of his early life are sketchy, it is clear that his father, Sheldon, was wealthy enough to provide him with a good education in New York, London and Paris. In his late twenties Edwin married, had children and for a while lived the life of a prosperous gentleman of leisure in New England. But then in his early thirties he became involved in a scandal of some kind and was forced to leave the United States. Whatever the cause of the row, it left him estranged from his family, down on his luck and cut off from funds. He was forced to earn a living as best he knew how. And what Smith knew best was Egypt. Ever since Napoleon's armies had swept down the Nile some fifty years earlier, the land of the pharaohs had fascinated the Western world. Egyptology had become a fashionable subject for serious academic study in Europe and America and tourists now added their graffiti to that carved by La Grande Armée on the monuments of the ancient kings. As a young man Smith had developed an interest in Egypt and read excitedly of the archaeological discoveries being made there. He had studied hieratic – the written language of pictograms and hieroglyphics used by the Ancient Egyptians that was being painfully deconstructed by professional scholars from clues in scraps of old papyrus and tablets like the Rosetta Stone. In fact he became something of an expert – albeit an amateur one – and when he was looking for a place to establish himself, far away from the miseries of his personal life, it was only natural that he turned to Egypt.

He settled in Luxor, the town built adjacent to the ancient ruins of Thebes in what had become known as the Valley of the Kings. He was the first American to live in Egypt and achieved a certain notoriety because of it. He had only £60 when he arrived in 1858 but he was a resourceful man (an Indiana Jones of his time) and soon found ways of turning a penny and advancing his archaeological interests. He put his small capital to two uses, becoming a money-lender (enjoying the lucrative interest rate of 5 per cent a month) and a dealer in antiquities. One fed the other – often the poor Egyptians to whom Smith lent money were the same people who sold him artefacts, and so he was able to get things at a very advantageous price. These he sold on to the tourists, collectors and Egyptologists who came through Luxor on their sightseeing trips down the Nile and who preferred to buy their souvenirs from an apparently trustworthy English-speaking expert.

As a result he gained a wide circle of friends and acquaintances. Among the expatriate community these included the great and the good; society people like Lucy Duff Gordon, the wife of a wealthy English baronet and friend to such literary celebrities as Dickens, Thackeray and Tennyson. She was living in Egypt for a time as the climate was said to be good for her health and compiling a book of her letters that went on to become one of the bestsellers of the Victorian era. Another intimate was Charles Goodwin, a well-known English Egyptologist based in Cairo with whom Smith kept up an intermittent correspondence about texts in the monuments at Luxor.

But he spent most of his time among the locals. One of his closest, if most turbulent, relationships was with his landlord, Mustapha Agha Ayat, an Ethiopian by birth and a merchant by profession. Agha had secured the posts of consular agent in Luxor to the British, Belgian and Russian governments and had his own flourishing antiquities trade on the side. He also had extensive property interests in the town and when Smith first arrived he had let him a house on a beautiful site adjacent to the temple of Rameses II. Over time they became friends, but it was a friendship often complicated by their rivalry in business. While they would sometimes work as partners, using each other's contacts to obtain and sell artefacts, neither was above seeking an advantage over the other when the opportunity presented itself. Outsiders were never quite sure which state of affairs was prevalent at any given moment.

But for all Edwin Smith's roguery, his genuine scholarship and his knowledge and interest in Egyptology are uncontested. If he was wont to sell a convincing fake or two, he was also able to spot the genuine article, and he treasured these artefacts when he found them. Thus it was that on 20 January 1862 he made one of the most important discoveries in the history of medicine – when he bought two tattered papyrus scrolls for £12.*

When he had a chance to examine his purchases more closely, the papyri turned out to be proto-medical textbooks; one describing forty-eight surgical cases, their diagnosis and treatment, the other a more extensive but haphazard collection of medical conditions and remedies. The former came to be known after its purchaser as the Edwin Smith Surgical Papyrus, the latter as the Ebers Papyrus (after a German professor who later acquired it from Smith). Both were old, very old, dating from around 1534 BC and written (probably by the same scribe) in a form of hieratic known as Middle Egyptian. But in content they were much older still. The papyri were copies of earlier documents dating back at least another 1,000 years and probably more. As James Breasted, an American Egyptologist, said of them later, it was as though someone today decided to sit down and copy out a manuscript written in the reign of Charlemagne.

Although both documents are remarkable because of what they reveal about the ancient Egyptians' practice of medicine, it is the Ebers Papyrus that occupies the central place in this story.

---

* Before they showed this papyrus to Smith, the grave-robbers stripped off some tattered remnants on the outside to make it more presentable. Two months later they glued these fragments to another relatively worthless scroll and sold it to him again. Smith detected the fraud and reconnected the genuine strips to the original papyrus, thus recovering for posterity a vital section relating to the heart.

At 110 pages, it is by far the longest and most comprehensive medical papyrus ever recovered and studied by Egyptologists. It is also very dense. As is often the case with such documents, there is writing on both sides; papyrus was expensive at the time and scribes who wrote to order for cost-conscious masters were loath to waste any of it. It is dated, on its reverse side, to the ninth year of the reign of Amenhotep I, which puts its physical origins at around 1534 BC – from the period known as the Middle Kingdom. However, references within the text suggest that it is actually a copy of a manuscript, probably written around at the time of the Old Kingdom, around 3,000 BC. Each page and each of the 877 paragraphs are numbered, which gives it the appearance of one cohesive document and explains why some people have called it the first book in history. But this is a little deceptive. In reality the papyrus is a loose assembly of different medical texts, put together in such a way as to suggest the author had drawn on a number of different sources and didn't quite understand how they should all fit together.

The texts deal with internal medicine (as opposed to external injury) and cover a huge range of conditions, everything from intestinal worms, eye complaints and ulcers to tumours, gynaecological disorders and heart disease. None of them, of course, is described as a doctor would today. While the document clearly shows that the ancient Egyptians had a rudimentary understanding of the circulatory system and some knowledge of basic anatomy, many of their superstitions, reasoning and treatments are based on concepts that are alien to us.

One of the most important of these is the notion of *wekhudu*. The Egyptian theory was that four elements flowed around the bodily system: blood, air, water and *wekhudu* – the corrosive leftovers of bodily waste. *Wekhudu* was thought to be the element most responsible for illness: too much of it in your system and you fell sick. Consequently, driving it out or dealing with its effects was a central tenet of ancient Egyptian medical practice and it's easy to see why purges and enemas were among the most common treatments used by physicians.

But as the Ebers Papyrus makes clear, they also had an extensive pharmacopoeia to draw on – around 160 herbal and vegetable-based remedies. Only around 20 per cent of these have ever been identified with certainty (some of the plants are now extinct or at any rate no longer growing in the Nile Valley and Egyptologists will argue for ever about the rest). However, among those that *are* known are many names familiar to us today – lotus, onion, watermelon, tamarisk, myrtle, juniper, cinnamon, date, dill, almond, celery and aniseed. How many of the 160 were truly effective as medicines is open to question. Some may have been, others may have been positively harmful. In many cases we just don't know. Less than 10 per cent of the world's current botanical species have ever been systematically investigated for any possible therapeutic properties, let alone plants from the dim and distant past.

But there is one species mentioned in the Ebers Papyrus that stands out, precisely because it *has* been systematically investigated. The ancient Egyptian name for it is *tjeret*. The Latin name for it is *salix*. We simply know it as willow – and willow is vital to this story because it contains the key ingredient of the most remarkable drug the world has ever seen: aspirin.

The true origins of willow's use as a medicine probably lie tens of thousands of years in the past – well before the Ancient Egyptians, indeed well before anything that can realistically be called a civilization took hold – when primitive man first got the notion that if he fell ill or was injured certain plants could help him recover.

How this idea developed is anyone's guess but most likely it evolved as Neanderthal man gave way to *homo sapiens* and man's reasoning powers and his awareness of the world around him slowly became more acute. It may have been instinctive at first, an intuitive response to maddening nausea or chronic pain – a sick person reaching out for the nearest plant in a desperate search for a palliative. Or it could have started with mimicry; born from the observation that a sick animal will seek out and eat a particular plant, or roll in a particular patch of grass. One can imagine how useful such knowledge would be to a hunter on the trail of a weak prey, and it could have been remembered and copied when the hunter himself fell ill. Either way, done enough times (and allowing for all the fatal experiments with poisonous plants that there must have been), many trees, herbs, roots and leaves with healing properties would slowly have been identified. Over the millennia, as man crept towards the earliest civilizations, he would have acquired enough knowledge to fill a basic medicine chest.

The willow tree would have been a useful addition to this primitive pharmacopoeia. For a start, it was widely available – there are over 300 known variants of the willow species and the tree grew in most parts of the prehistoric world. More importantly, it contained alkaloids (called salicylates) that under certain circumstances can reduce fever and relieve pain in humans.* This would not be known or understood for several thousand years, of course, and it is probable that the primitive pharmacist was attracted to the willow only because of its bitter taste, or because he liked the shape of its leaves, or because it grew in a certain place. But repeated use would have released some of its secrets.

The very earliest known written reference to willow's medicinal properties has been found on a stone tablet from the third dynasty of the kings of Ur. This was one of the city-states of the Sumerian civilization that sprang up on the fertile plain between the Tigris and the Euphrates rivers around 5000 BC. There has been much weighty academic argument over the years as to whether or not Sumerian medicine amounted to very much. It was bound up with magic and spells and the casting out of demons and while it was presumably considered effective by the patient there is no real way of knowing whether it had any genuine physical benefits. But the Ur III tablet, which dates from around 3000 BC, does suggest that Sumerian medicine had its practical side. It contains a list of more than a dozen prescriptions using raw materials like turtle shell, snake skin and milk, along with plants like myrtle, thyme, fig, date and willow. Unfortunately, there is no corresponding list of the diseases for which these remedies were intended. And that's where the Ebers Papyrus comes in.

The Sumerian city-states were on the wane when the first Egyptian civilization was becoming powerful, but for a time they co-existed and there were contacts between the two. As early as the third millennium BC there were simple trade links all around the eastern shores of the Mediterranean and up and down the Mesopotamian deltas. Where trade flowed, people and knowledge followed. New ideas would have passed from one to the other and conclusions independently arrived at would have been reinforced and validated by the experience and influence of

* Botanists have a number of theories as to why the salix species contains these chemicals. The most important of these is that they may help the plant fight off contagion, initiating a kind of cell suicide process known as apoptosis, in which a diseased leaf will die and fall off to avoid infecting the others. It has also been suggested that the willow tree excretes salicylates to deter insect predators.

outsiders. This applied to medical knowledge as much as anything else, and while there is no direct connection between the Ur III tablet and the original Ebers Papyrus (beyond the fact that they were both written at roughly the same time), it is likely that many of the prescriptions and remedies they have in common were designed with the same illnesses in mind. It follows, then, that the Sumerians and Egyptians could have used willow for roughly the same things.

It is not easy to pin down exactly what those were. Although the Ebers Papyrus does relate its remedies to diseases, many of them have no direct modern equivalents and are not easily translatable. Egyptologists have argued for years over some of Ebers's terms and because many of its underlying medical concepts are alien to us they are subject to different interpretations. None the less, the three clear references to willow in Ebers all point towards the plant's use as either a general purpose tonic or as some kind of pain reliever and anti-inflammatory therapy.

The first of these comes in a section sandwiched between remedies for the regulation of urine and those for treating a cough. Willow is one of the ingredients – others are figs, beer and dates – in a treatment taken orally to 'cause the heart to receive bread'. Again, no one is quite sure what this phrase means but it has been suggested that it refers to a general pick-me-up remedy, a kind of medicinal drink for non-specific aches and pains. Presumably the willow ingredient would have come from a dried and ground-up version of the bark or leaves, which was then added to the fluid.

The next two references are to willow's use as an ingredient in externally applied salves – for ear infection and for making the *met* supple. The *met* is thought to refer to the body's muscles and tendons and it's possible that willow was used to treat them when they became stiff or inflamed, perhaps because of conditions like arthritis. Ear infection can be painful too and willow's analgesic properties might have been useful in relieving the pain.

In all three of these remedies willow would seem to have been the only active ingredient. None of the others in the various recipes – caraway seeds, figs, dates, beer and lotus leaves – is known to have any specific medical value other than the purely nutritional.

But were any of the remedies effective? It's hard to say – particularly in the case of the externally applied salves, because patients would have had to rub vast quantities of the stuff into their skin in order to absorb enough useful salicylates. But maybe that's what they did. We just don't know for sure. If it *did* work, then willow would certainly have filled a gap in the physician's armoury. At this stage in their development the Ancient Egyptians had few other pain relievers to hand, or at least none they regularly made use of. They didn't yet use opiates for example; although they knew of plants like the poppy and the mandrake, they were ignorant of their effects on the central nervous system and of their potential as sedatives. Neither is mentioned in Ebers. They did know of cannabis, however, but seemed to use it only as a salve and emetic. They certainly did not appreciate its narcotic effects. Indeed, their most widely available painkiller seems to have been alcohol; there are various beer- and wine-based remedies quoted in the document, and getting the patient ceremonially drunk seems to have been a favoured solution to quite a few of the problems Egyptian physicians faced.

Willow's case is given a further boost by the fact that the Ancient Egyptians also used myrtle – another salicylate-bearing tree – in a very specific remedy to treat rheumatism in pregnant women. A decoction of dried myrtle leaves would be steeped in 'the dregs of excellent beer' and then applied to the patient's abdomen and back. What effect this would have had is unclear, but interestingly, both myrtle and willow would be used in Europe many centuries later as a rheumatism treatment.*

The identity of the author of Ebers is a mystery (the papyrus bears no signature) but it would have been a valuable and much prized document at the time. Edwin Smith was told it was found with a mummy in the Assassif area of Thebes, a highly prestigious burial site. If the story is true, then the papyrus probably belonged to a very high-ranking doctor indeed. The medical profession, held in great esteem in Ancient Egypt, was highly stratified, from the simple practitioner at the bottom right up to the chief palace physician. The more senior the doctor, the better located and the more elaborate and well equipped his tomb would have been, in order that he might carry on the good work in the next

* Other trees and plants that contain salicylates include wintergreen, black cohosh root, poplar tree bark, and sweet birch bark. The concentration in willow is actually the lowest of all these plants.

world. Anyone buried in such a place as Assassif and with such a remarkable set of papyri as Ebers and the Edwin Smith Surgical Papyrus would have been held in the highest regard by his peers.

Of course, the status and respect accorded to these eminent practitioners and their knowledge lasted far longer than the power of the various dynasties they served. Even when other expanding empires – Ptolemaic, Persian, Greek, Roman – began to whittle away at Egyptian independence, the skills, experience and remedies of Egyptian physicians were much in demand and much imitated. Just as they had learnt from the Sumerian civilization, so succeeding civilizations learnt from them. Through trade, military contacts, and the links established through coastal cities like Alexandria, their influence and wisdom would be handed down from generation to generation and go on to inform the development of medicine throughout the Mediterranean.

This may explain why, more than 1,000 years after the Ebers Papyrus and its secrets were sealed up in a tomb at Thebes, Classical Greek physicians were still dispensing remedies very similar to those the document had described, and why willow was among them.

The most famous and important of these Greek physicians was Hippocrates, 'the father of medicine', who, with his many followers, began to free the profession of healing from the shackles of mysticism that had chained it for so long. Until then the relief of pain had been closely associated with spiritual belief, the idea that supernatural causes were as likely to be responsible for human suffering as physical ones. The doctors of Ancient Egypt who drew up and referred to documents like the Ebers Papyrus may have sought rational responses to pain, but they were trained in a temple-like setting and were as much priests as they were physicians.

But from Hippocrates, who lived and practised on the Greek island of Cos in the fifth century BC, came the first glimmerings of understanding that by careful, painstaking observation the doctor became a servant of nature rather than a sorcerer. This led to the Hippocratic Corpus, a huge body of medical texts (contributed to by a number of different physicians) that listed diseases, diagnoses and treatments in terms that are free of mumbo-jumbo and familiar to doctors today. No one is absolutely sure whether those treatments, particularly the plant-based ones, were independently arrived at as a consequence of this new deductive reasoning or were embellishments of ideas inherited from the past. But the principle of using plants as medicines had certainly been handed down and many of the remedies would have been just as recognizable to healers 1,000 years earlier.

In any event, one thing is beyond doubt. Much as the Ebers Papyrus had done, Hippocrates recommended the bark of the willow tree as an analgesic, although in his case it was as a remedy for the pains of childbirth and as a fever reducer. It was, of course, only one of hundreds of different ingredients in hundreds of different treatments and remedies – just another useful plant remedy with nothing too remarkable about it. But his gentle plug placed willow firmly into the pharmacopoeia of the Classical era and there it was to stay for hundreds of years.

We know for instance that in AD 30 the Roman physician Celsus used extracts of willow leaves to relieve what he described as the four classic signs of inflammation (*rubor*, *calor*, *dolor* and *tumor*, or redness, heat, pain and swelling). Not long afterwards, Pedanius Dioscorides, a Greek doctor who served as a botanist in the armies of the Emperor Nero, wrote of the therapeutic potential of willow in his *De Materia Medica*, a work that survived to be translated by the Arabs. So too did Pliny the Elder, the Roman general who completed his thirty-seven-volume *Natural History* in AD 77 (just before his death in the ashes of an erupting Mount Vesuvius). And Claudius Galen, who studied in Egypt and worked in Greece as a physician to the gladiators before becoming physician to the Emperor Marcus Aurelius, recommended willow as a treatment for mild to moderate pain. In fact, by the time Galen died in AD 216 willow was a common remedy across the civilized world.

But it was a world that was soon to disintegrate into the darkness and ignorance of more brutal times. With it went much of the accumulated medical wisdom of thousands of years and many of the remedies, like willow, which had been in use for so long. Although other cultures would continue to find their own medical uses for the plant, it would not reappear properly until the eighteenth century, when it took a medically minded country parson to rediscover something the Ancient Egyptians had known several millennia earlier. Only then would it start living up to its heritage.

As for Edwin Smith, the papyri he found did little to improve his fortunes, although it is clear he valued them. In 1864, he wrote to his friend Charles Goodwin: 'I do not wish to sell either Papyrus. They are part of my collection and the prices put on them (as I profess to sell antiquities) are intended to be prohibitive.'

Sadly, by 1869 Smith had suffered a number of reverses, including a distressing spell of temporary blindness, and money was very tight. In the late summer of that year, a sales catalogue circulated among buyers of Egyptian antiquities. It included an advertisement for 'a large medical papyrus in the possession of Edwin Smith, an American farmer [sic] of Luxor near Thebes'. The agent for the sale was none other than Mustapha Agha, Smith's old friend and sparring partner.

It was bought for an undisclosed sum by Georg Ebers, a German professor of Egyptology and author of a series of popular historical works about the pharaohs. To his credit he had a good stab at translating it into German, but then rather tarnished his reputation by shabbily pretending to have been the one to find it as well – probably through eagerness to have his name associated with the document (papyri were often named after the person who discovered them). In any event it worked. The Ebers Papyrus was first published as a facsimile in 1875 and later passed to Leiden University where it still resides today.

A few years after its sale Smith left his home among the tombs, either as the result of a partial reconciliation with his family or, more likely, because he had run out of gullible tourists to whom he could sell *antikas*. Whatever the case, he departed from Egypt and then disappeared from view. On his death in Naples in 1906 his daughter Leonora donated his remaining treasure, the Edwin Smith Surgical Papyrus, to the New York Historical Society. Smith's name lives on in that one eponymous text, but, sadly for a man who brought to light two of the most important documents in the history of civilization, his intriguing character has been largely forgotten. Without him, the origins of the world's most remarkable medicine might never have been known.

2

THE BARK OF AN ENGLISH TREE

My Lord,

Among the many useful discoveries which this age has made, there are very few which better deserve the attention of the public than what I am going to lay before your Lordship.

The Reverend Stone paused for a moment and gazed at these words. He was making a bold claim and he knew that the man to whom his letter was addressed would raise his eyebrows when he saw it. The Right Honourable George, Earl of Macclesfield was nobody's fool – he wouldn't have been President of the Royal Society if he were – and no doubt madmen and charlatans made similarly presumptuous declarations to him all the time. Stone could trade on a slim acquaintance with His Lordship and was counting on that to ensure that his letter was read, but when all was said and done, he was still only a country parson. He would have to be careful not to overdo it.

He took up his quill, refreshed its point in the inkwell and scratched on.

There is a bark of an English tree, which I have found by experience to be a powerful astringent, and very efficacious in curing agues and intermitting disorders. About six years ago, I accidentally tasted it, and was surprised at its extraordinary bitterness; which immediately raised in me a suspicion of its having the properties of the Peruvian bark. As this tree delights in a moist or wet soil, where agues chiefly abound, the general maxim, that many natural maladies carry their cures along with them, or that their remedies lie not far from their causes, was so very apposite to this particular case, that I could not help applying it; and that this might be the intention of Providence here I must own had some little weight with me.

As his spidery handwriting filled the page, his thoughts drifted back to that moment, five years before, when Providence had guided his actions. It was market day, he recalled, and he had gone out for a stroll to relieve the stiffness in his joints . . .

Chipping Norton was a sizeable place by eighteenth-century standards and was usually pretty busy. But it was especially so on that mid-summer morning in 1758. The warm weather had brought in the crowds from nearby towns and villages; landowners and labourers, travelling tradesmen, a few fine ladies with their attendant servants and even one or two black-clad scholars from nearby Oxford, enjoying a day away from their gloomy labours. The townspeople welcomed them all – even the poorest might have money to spend – and with the shouts of stallholders echoing above the bleating and bellowing livestock, a pleasant holiday atmosphere prevailed.

To many of those in the crowd the Reverend Edward Stone would have been a familiar if unconventional figure. Chipping Norton wasn't his parish, but he'd lived on the outskirts of the town for twelve years and on market days he made a point of going out among his fellow citizens. Perhaps he stopped to chat with friends or exchange pleasantries with the market traders, the warm sunshine easing his rheumatism and improving his mood. He was reasonably popular with his neighbours, respected rather than liked – there was a fussy and meticulous side to his character that discouraged intimacy. But his Whig politics in what was at heart a Whig town would have made him more approachable than the town's own vicar, a staunch and unbending Tory.* Many must have wished that he held the parish instead.

But Stone had other commitments, more interesting and less strenuous than those offered by the dull, busy life he'd have had as parson to the people of Chipping Norton. Some years before, he'd secured an enviable living as chaplain to the family of Sir Jonathan Cope at their

* During a famous election in 1754, when the rural Oxfordshire seat was won by the Whigs, the Tory vicar of Chipping Norton barricaded himself into the church to stop townspeople getting in to ring the bells in celebration.

seat of Bruern, eight miles away.* His duties there were modest – a brief service for the household every few weeks and occasional prayers, blessings, weddings and funerals when necessary. At other times he may have been required to join his employer in philosophical conversation as they wandered around the ruins of the old Cistercian priory in Bruern's grounds, or to make up the numbers at dinner when the Copes had guests, but it wasn't an onerous life. And although the eight-mile journey on horseback had its disadvantages (it was uncomfortable in winter and footpads and highwaymen were an ever-present threat), Stone had grown used to it. Even the small stipend was not a concern; he had a reasonable private income of his own, supplemented by the rectorships of two other nearby parishes at Horsenden and Drayton. He had managed to sub-contract those duties to curates and was rarely called on himself.

But the biggest advantage Stone had gained from his association with the Copes was the time and freedom he had to pursue his many other interests. He was a Justice of the Peace (duties that mostly involved the administration of the Poor Law, expected of someone in his position and in his case enacted with meticulous fairness), and of course he had his politics. From his college days he had also retained an interest in theological, mathematical and astronomical subjects. He had come a long way in his fifty-six years.

The only son of a yeoman of modest means, Edward Stone was born on a farm near Princes Risborough, Buckinghamshire in 1702. There is no record of his early life, but somewhere along the way – perhaps because it offered possibilities for advancement that would otherwise have been denied to someone from his background – it was decided he should enter the Church. In those days that meant a university education, and in 1720 he won entrance to Wadham College, Oxford. He graduated four years later, gained his MA in 1727, and ordination and a brief period as vicar to the people of Charlton-on-Otmoor followed. But in 1730 he returned to Wadham as a fellow and for the

* The Copes were a well-established landowning and military family. Sir Jonathan's father had earned a footnote in history some years earlier when Bonnie Prince Charlie defeated him in battle.

next eleven years held posts as the college librarian, bursar, dean and sub-warden.*

Stone's fortunes only improved from then on, with marriage to Elizabeth Grubb, the daughter of a wealthy Buckinghamshire landowner, and the pleasant livings at Horsenden, Drayton and Bruern. When he finally arrived at Chipping Norton in 1745 he was wealthy enough to acquire one of its more impressive properties.

> Two handsome parlours, storeroom, kitchen, excellent vaulted cellars, four bed chambers and four good garrets, offices, brewhouse, coalhouse, stables for five horses, dairy – with apartments over them, a garden, two acres of adjoining pasture and twelve acres at a small distance.

To those twelve acres he had added another ten over the years and it was his custom on fine days to stroll out from his house at the top of the hill on which Chipping Norton is set, and down to his land on the opposite side of town.

As he made his way through the marketplace, past the auctioneer, the sheep pens and the wool buyers, he would have passed a number of simple stalls. Most sold the basic foodstuffs that the townspeople needed, but there were others with more alluring goods – cloth and lace, ironmongery, sweetmeats and remedies. As usual, the last of these would have been the busiest, surrounded by a small crowd listening intently to the claims of the hawker. Health was a serious matter in those days and everyone dreaded falling ill.

In common with other parts of Europe and North America, eighteenth-century Britain was slowly gearing up for a medical revolution.

---

* An interesting insight into Stone's character can be gained from his involvement in an infamous scandal that broke at Wadham during his time there. A commoner (the lowliest rank of scholar) called William French complained to Stone that the Warden of the college, one Robert Thistlethwayte, had sexually assaulted him. To his credit, Stone didn't turn him away and hush the matter up as others might have done, but bravely helped initiate proceedings against Thistlethwayte, his ultimate superior in the college hierarchy. Thistlethwayte had to flee the country, leaving contemporary satirists to make play with the unfortunate alliterative potential of the name of the college and the crime of which he had been accused. Thus: *There once was a Warden of Wadham/Who approved of the folkways of Sodom/For a man might, he said/Have a very poor head/But be a fine fellow at bottom.*

Most people didn't know it yet and the signs were barely visible, but things were beginning to happen: a few new hospitals in the larger towns and cities (eighteen by 1752), an important new medical school established in Edinburgh, knowledge of such things as anatomy, circulation and the nervous system growing all the time. There were even glimmers of understanding about infection and the science of pathology. Over the next 150 years, as industrialization wrought its massive changes on society, these small seeds would flower and prosper as doctors and scientists unlocked the secrets of hygiene, antisepsis, anaesthesia, vaccination, germ theory, nutrition, and much else. Individual and public health would be slowly and painfully transformed.

But all that lay in the future. The day-to-day reality for Edward Stone and for most of his contemporaries was a system of medical provision (if one can actually call it a system) that hadn't changed much since the Middle Ages. Surgery, for instance, the poor relation of the medical profession, had only recently managed to assert its independence from the barbering trade (a separate Company of Surgeons was established in 1745) and was still largely the practice of robust manual feats like bone-setting, amputation and the removal of gallstones, whereas the average physician of the time – though he enjoyed rather more status, earned a little more money and dressed his diagnoses up in a better class of Latin – continued to cling conservatively to a curious hotchpotch of theories and treatments that hadn't advanced significantly since Classical times.

The most influential of these was the Hippocratic belief in the influence on illness of the body's Four Humours (blood, phlegm, black bile and yellow bile). According to this theory, further developed by men like Claudius Galen and resurrected by medieval and Renaissance scholars, the humours were the main cause of disease. To keep them in balance and thereby to treat an illness effectively, a doctor either gave fluids to his patients or drained them away: letting the liquids out by bloodletting, leeches, enemas, purges and emetics, and adding them by poultices, salves and orally taken botanical and mineral remedies. Many of the latter were useless or, even if sometimes effective, used inappropriately. Some, like the administration of poisonous mercury for syphilis and open sores, could be downright lethal. Putting yourself in the hands of a mid-eighteenth-century doctor was like buying a lottery ticket. If you were lucky and found a practitioner with common sense, someone who valued bed rest and good nursing over the more savage procedures of his profession, then you might survive. If not, then as a contemporary satirist of the day, Matthew Prior, once quipped, 'Cur'd yesterday of my disease, I died last night of my Physician.'

It is not surprising that these traditionalists were in the majority. Physicians had little desire or incentive to keep up with new developments once they had received their initial training. Such small scientific advances as there were took place in a few centres of medical learning like London and Edinburgh, and there was no formal system for disseminating this knowledge around the country. New ideas and new treatments could take years to seep out to the general practitioner and even when they did, few had the patience or understanding to try them out.

Not that this lack of up-to-date knowledge made much difference to rural inhabitants, of course, because few of them ever saw a doctor. The majority of qualified practitioners preferred to spend their time with their well-to-do patients in the larger towns and cities. If anyone in Chipping Norton, for example, wished to see a doctor, they had to send to Oxford for one to come out in his carriage. That was expensive and well beyond the reach of most.

So people fell back on the only other alternatives: self-medication or the services of the local apothecary. There was little difference between them. Most of the brews and specifics that the latter dispensed were dressed-up versions of the same herbal folk remedies that everyone practised at home and for the most part just as ineffectual.

Which was why on market day, a stall that sold cures with such impressive titles as Aqua Mirabilis, Elixir Vitae or Heavenly Potion always did good business. And even if they didn't live up to the pedlar's promises and cure your chronic back pain or make your boils disappear, most of them contained sufficient alcohol to induce a temporary sense of well-being. In any case, the hawkers had a good line in health patter and you never knew what useful medical tips you might pick up.

The Reverend Stone would have been as aware as anyone else of the discomfort and dangers of falling ill. He was fifty-six, already well past the average life expectancy for the eighteenth century, and he suffered from the same occasional fevers and rheumatism that afflicted anyone of his age. He was a man of science and of the cloth, and therefore supposed to be above such things, but he was a man of his time too and must have been as susceptible as his neighbour to a remedy that promised to cure his stiff joints, or would at least have been tempted to stop and listen to the descriptions of what it could do. Not on this day though. Today he walked on.

His route took him across Chipping Norton's main street, past the White Hart coaching inn and down towards the fields that lay on the north-east side of town. He crossed some scrubby common land, grazed by a few cows and sheep, and then came to his own property. This had been enclosed only a few years earlier and a small stream known as the Common Brook ran right through it. Before he had bought the fields, the town authorities had planted willow trees along the bank of the stream and they were now mature enough to offer shade and a pleasant place to sit and think.

It was there he had his revelation.

We don't know what prompted Stone to take a piece of willow bark and put it in his mouth. It might have been no more than idle curiosity. But whatever it was sparked off a remarkable train of thought. As the bark touched his tongue he was struck by its acute bitterness, a strangely familiar taste. He sat there for a while, lips pursed at the sharpness of it, trying to remember what it might be. Then it came to him. It was almost exactly like that of a remedy that doctors used to treat the agues.

The agues were a significant medical problem in parts of Britain, continental Europe and the New World. Sometimes fatal, always unpleasant, they were usually categorized according to their intermittent and seasonal appearance with Latin names such as *quotidian, tertian* and *quartan*, and had been a recognizable and much puzzled-over condition for many hundreds of years. Striking indiscriminately, without deference to rank or fortune, attacks sometimes reached epidemic proportions, burning their mark on the consciousness of successive generations. In the fourteenth century Geoffrey Chaucer, in 'The Nun's Priest's Tale', wrote, 'You are so very choleric of complexion. Beware the mounting sun and all dejection. Nor get yourself with sudden humours hot. For if you do, I dare well lay a groat. That you shall have the tertian fever's pain. Or some ague that may well be your bane.' Two hundred years later Shakespeare would mention the agues in eight of his plays and they would remain a mysterious but potent threat to good health in Britain up until the late nineteenth century.

In 1769, a Scottish physician named William Buchan had a go at describing their causes and symptoms in his book *Domestic Medicine*.

> Agues are occasioned by effluvia from putrid stagnating water. This is evident from their abounding in rainy seasons and being most frequent in countries where the soil is marshy, as in Holland, the Fens of Cambridgeshire, and the Hundreds of Essex. This disease may also be occasioned by eating too much stone fruit, by a poor watery diet, damp houses, evening dews, lying upon the damp ground, watching, fatigue, depressing passions and the like . . . An intermitting fever generally begins with a pain of the head and loins, weariness of the limbs. Coldness of the extremities, stretching, yawning, with sometimes great sickness and vomiting; to which succeed shivering and violent shaking. Afterwards the skin becomes moist, and a profuse sweat breaks out, which generally terminates the fit or paroxysm. Sometimes indeed the disease comes on suddenly, when the person thinks himself in perfect health; but it is more commonly preceded by listlessness, loss of appetite and the symptoms mentioned above.*

This reflected the prevalent medical thinking about the disease, although the name ague was also mistakenly used to identify a variety of conditions, from influenza to migraine, that had broadly similar symptoms, the limited eighteenth-century medical vocabulary having to do duty for the thousands of conditions and complaints we know of today. In actual fact it should have only referred to one: malaria.

There were five species of the *Anopheles* mosquito in eighteenth-century England, which bred wherever there were marshy or brackish waters.† The female of each species was capable of carrying the protozoan plasmodium parasite in its stomach and passing it to humans

---

* Dr William Buchan was an eighteenth-century success story. His book became one of the bestsellers of the age, particularly in postcolonial America where domestic medicine was a fixture in every middle-class home. On his death in 1805, he was buried in Westminster Abbey.

† The most competent being the *Anopheles antroparvus*, which preferred to breed in river estuaries. Although all five species can still be found in this country, they are no longer malarial here. However, some scientists have warned that global warming could one day see the return of the disease.

when it penetrated the skin and sucked blood. The parasite was the cause of malaria, for so long the curse of tropical countries, and, so it has subsequently been discovered, of England's ague too. None of this was understood by eighteenth-century physicians, of course, but it was known that the ague most commonly occurred in areas close to dank or stagnant water. For centuries doctors had believed that the disease originated in the odours or miasmas given off by these waters. It wasn't until 1897 that anyone thought to raise their eyes from the waters to the swarms of biting insects above them.*

The bitter-tasting remedy that Edward Stone had recalled was the bark of the cinchona tree, or Peruvian bark, as he would call it in his letter to the Earl of Macclesfield. It got its name from its origins in the New World. When the Spanish conquistadors began their conquest of Central America, they found it as afflicted by agues and fevers as the world they had left behind. But they also found a remedy. It was first described by a monk, Father Antonio de la Calaucha, in the *Chronicle of St Augustine* in 1633: 'A tree grows which they call the fever tree in the country of Loxa whose bark, the colour of cinnamon, made into a powder amounting to the size of two small silver coins and given as a beverage, cures the fevers and tertiana. It has produced miraculous results in Lima.'

The Spanish called the remedy *quina*, after its Peruvian name of *kina* (many years later the active alkaloid it contained would be called quinine, a derivation of both). Whether the indigenous population had used it prior to the arrival of the conquistadors isn't known, but according to popular (and possibly inaccurate) legend its reputation was supposedly made when it was given to the wife of the Viceroy of Peru who had contracted malaria. She recovered, and it was soon being exported to Europe. Pope Innocent X asked a Jesuit priest, Juan de Lugo, to test it, and shortly afterwards, in the late 1640s, directions for its use were disseminated by his order throughout the continent.

It wasn't universally adopted. The bark sometimes didn't work. It couldn't cure all fevers, only those associated with malaria, and as the term ague was often a broad description for a range of unrelated

* In 1880, a French physician, Alphonse Laveran, found the parasite in malaria-infected blood cells. In 1897, an Englishman, Sir Ronald Ross, working in India, deduced that they were transmitted by mosquitoes.

problems, a remedy that promised to cure them all could sometimes be ineffective. It also took a while, particularly in Protestant England, for people to overcome their suspicion of a remedy promoted by Jesuits. This prejudice was so deeply held that it was said Oliver Cromwell had died of ague/malaria rather than take Jesuit's Bark. A humble English apothecary called Robert Talbor was able to exploit these fears when he developed his own secret remedy for the agues and made a big selling point of the claim that it had nothing to do with the Jesuits. The success of his treatments made him famous and he was appointed Physician Royal to Charles II in 1672. He was knighted a little later when he cured the king of a fever and was soon on his travels around Europe dispensing the remedy to such celebrities as the son of Louis XIV of France and Louisa Maria, Queen of Spain. On his death his secret remedy was revealed to be . . . Jesuit's Bark, the bitter taste of which Talbor had merely disguised by infusing it in strong white wine. Once the news was out, cinchona became more widely accepted. But as the tree grew only in Central America and the Spanish monopolized the trade, it was often unavailable and always very expensive.

A cheaper domestic equivalent, if such a thing existed, would be a very significant discovery indeed.

How much of this ran through the Reverend Edward Stone's mind as he sat beside his willow trees in rural Oxfordshire? He had just tasted something that reminded him strongly of the one effective remedy for the ague, which everyone knew was caused by damp vapours. Much like those, in fact, that could be found along his own stretch of the Common Brook.*

It was then that he made his next intuitive leap and remembered

---

* The trees are still there – or at least their botanical descendants are. I set off to find them one day, armed with a copy of an eighteenth-century map that detailed the boundaries of enclosed land around Chipping Norton and a more modern one of the area from the Ordnance Survey. The common is still there, although one corner of it is now dominated by a huge building, built in the nineteenth century as a textile mill and now turned into luxury flats, but the brook runs through it on to what was once Reverend Stone's land. I traced the course of this for about a mile until I judged I was at the right spot and there found a dozen beautiful and very old willow trees. It was a warm summer's morning, pretty much as it must have been in 1757, and apart from two or three cows

the doctrine of signatures. As a former librarian of Wadham College, Stone would have had ample time to browse through its collection of books by eminent European thinkers. It was probably there that he first came across the work of Paracelsus, a Swiss botanist, natural philosopher and iconoclastic medical theorist, who both infuriated and enlightened the world of Renaissance medicine.* Paracelsus was convinced that nature was sovereign and only by knowing her and obeying her could a physician perform effectively. He drew heavily on a popular folk theory of the time, which he called the doctrine of signatures, to draw up remedies. According to the doctrine, nature gives signs to the close observer of the healing properties of certain herbs and trees. Hence the orchid could be used to treat venereal disease because it looks like a testicle, or the blue eyebright plant could be used to treat eye disease. It also followed that the cause of a disease could often be found close to the location where it first appeared, and that the cure would be present at the same place. This was why if you were stung by a nettle, the first thing you did was look around for the dock leaves that were sure to be growing nearby.

To Stone, there were compelling connections between water, the known cause of ague, and a possible cure for the disease. Willows grew by water and, as he had discovered, willow bark tasted like the bark of the cinchona tree. According to the doctrine of signatures, the case for willow's potential as an ague cure was already half made. But would it work in practice?

In great excitement Stone began gathering up twigs that had fallen from the trees and took out his pocketknife to slice some bark from the trunks around him. When he had an armful, he set off across his property towards a mill that sat in the distance at the other end of the

---

munching quietly among the thistles, there was no one else about. I sat down in the shade of the trees and nibbled on a piece of the bark. It was just as bitter as Stone had said it was. To my surprise there was nothing else to mark the place where this monumental discovery took place, which is something the people of Chipping Norton might want to remedy. A small plaque would suffice.

* His real name was Theophrastus Philippus Aureolus Bombastus von Hohenheim. He gave himself the name Paracelsus to denote the fact that he came after Celsus, the famous Roman doctor, and therefore represented a new way of thinking.

Common Brook. If the similar taste of his willows and cinchona had any significance, then he should experiment when the bark had been dried and preserved in the same way as the ague cure. The miller, a man called William Kench, had a large oven where he sometimes baked bread for the local farmers' wives.

What Kench had to say when the vicar arrived clutching an armful of twigs and made his odd request is not recorded. But he agreed to put Stone's bundle in a bag on the top of his oven and to watch over it to make sure it didn't burn. Over the next few weeks Stone made a number of visits to the mill to see how the willow bark was drying and to add to the pile. In between times, he began to scour the libraries at Bruern and Wadham College.

> It seemed probable, that if there was any considerable virtue in this bark, it must have been discovered from its plenty. My curiosity prompted me to look into the dispensatories and books of botany, and examine what they said concerning it; but there it existed only by name. I could not find, that it has or ever had, any place in pharmacy, or such qualities as I suspected ascribed to it by the botanists . . .

In the Western and Arab medical traditions, willow's therapeutic reputation had dimmed considerably since Classical times, becoming just another name in a huge list of plants that were considered (or had once been considered) to have some medicinal properties.* Over the centuries, as botanists categorized more flora, and apothecaries, monastic healers, herbalists and physicians experimented more widely, this list grew longer and longer and it was inevitable that many of the plants on it were either neglected or forgotten. Willow had been used as a folk medicine in parts of Europe throughout the Middle Ages and up to the Renaissance but it had gradually become more prized as a building material than a remedy. Even when it rated a rare mention in the

* Other cultures had independently learned to use willow medicinally. It appears in texts of Chinese medicine as early as the sixth century, although it is not clear how widely it was used or what it treated. There's also some evidence that it was used by indigenous tribes in southern Africa (by the Hottentots in particular – see Chapter 3) and by Amerindians prior to the arrival of Europeans, although as neither of these cultures had a written medical tradition, it is difficult to be specific about dates.

medico-botanical literature, there was confusion over how it could be used.

The best example of this is in a book by Nicholas Culpeper, called *The English Physician: or an astrologo-physical discourse of the vulgar herbs of this nation,* which had been written a century before Edward Stone's experiment. Culpeper is a legendary figure in the field of herbal medicine because of the way that he took on the medical establishment and challenged its dogma that health care was a matter for professionals. An apothecary's apprentice who set up as a doctor to the poor of London, he outraged the College of Physicians by translating its Pharmacopoeia from Latin into English, thus breaking its monopoly. *The English Physician* was built on this translation and contained over 500 plant-based prescriptions for various maladies. No doubt some of them were effective, but Culpeper's suggestions for willow's use show how far knowledge of its medicinal properties had declined since Classical times. His recommendations included using the bark as a treatment for staunching bleeding of wounds, reducing indigestion, replenishing dimmed sight, improving the flow of urine, treating warts and dandruff and staying the heat of lust in men and women. It is an impressive list, but sadly, mostly nonsense.*

So Stone had to conduct his own experiments. Determined to make an accurate comparison with the curative powers of Peruvian bark, he returned to Kench's mill to retrieve his bag of dried willow and took it home to grind it with a pestle and mortar. Once he had pounded and sifted about one pound's weight of it, he began looking around for ague sufferers.

> It was not long before I had an opportunity of making a trial of it; but, being an entire stranger to its nature, I gave it in very small quantities, I think it was about twenty grains of the powder at a dose, and repeated it every four hours between the fits; but with great caution and the strictest attention to its effects: the fits were considerably abated but did not entirely cease. Not perceiving the least ill consequences, I grew bolder with it, and in a few days

---

* On the contrary, as we have known for some time, it improves blood flow and upsets stomachs. But the idea that it can treat warts may have some merit, or so it is thought in parts of Italy, where one 'traditional' home remedy for a wart is to tape an aspirin to it for a couple of weeks.

increased the dose to two scruples, and the ague was soon removed. It was then given to several others with the same success, but I found it better answered the intention, when a dram of it was taken every four hours in the intervals of the paroxysms.

Although Stone does not identify the first recipients of the treatment, it is likely that they were either members of his own family and household or among the sick and destitute of Chipping Norton, none of whom would be in a position to argue with him about his right to use them as guinea pigs for his experiments.* But as the new medicine showed signs of being effective, word would soon have got around and more influential people would have begun to ask for his help. Almost certainly, Stone told his patron, Sir Jonathan Cope, about it and obtained his permission to treat the servants and workers on the Bruern estate. He may even have treated members of Cope's own family, the ague being no respecter of rank. As time went by, and the remedy (with a little modification) appeared to work against even the most obdurate cases, he became satisfied that his hunch was paying off.

I have continued to use it as a remedy for agues and intermitting disorders for five years successively and successfully. It has been given I believe to fifty persons, and never failed in the cure, except in a few autumnal and quartan agues, with which the patients had been long and severely afflicted; these it reduced in a great degree, but did not wholly take them off; the patient, at the usual time for the return of his fit, felt some smattering of his distemper, which the incessant repetition of these powders could not conquer. It seemed as if their power could reach thus far and no further, and I did suppose that it would have not long continued to reach so far and that the distemper would have soon returned with its pristine violence, but I did not stay to see the issue. I added one-fifth part of the Peruvian bark and with this small auxiliary it totally routed its adversary . . .

. . . By all that I can judge from five years experience of it upon a number of persons, it appears to be a powerful absorbent, astringent and febrifuge in intermittent cases; of the same nature and kind with the Peruvian bark and to have all its properties, though perhaps not always in the same degree. It seems, likewise, to have this additional quality viz., to be a safe medicine –

* There was no resident doctor in Chipping Norton at the time to contradict him.

for I could never perceive the least ill-effects from it, though it had always been given without any preparation of the patient.

After five years of successful experiments, Stone wanted to share his remarkable discovery. It is no surprise that he decided to write to the President of the Royal Society: its statutes might have been designed with him in mind:

> The business of the Society in their Ordinary Meetings shall be to order, take account, consider and discourse of philosophical experiments and observations; to read, hear, and discourse upon letters, reports and other papers concerning philosophical matters; as also to view, and discourse upon rarities of nature and art; and thereupon to consider, what may be deduced from them, or any of them; and how far they or any of them, may be improved for use or discovery . . .

The Royal Society originated in casual meetings between men of science and learning in London in the mid-seventeenth century, which gradually became formal weekly gatherings. At first a nameless and unofficial body, it gained a Royal Charter from King Charles II in 1663 as The Royal Society of London for Improving Natural Knowledge. Two years later, the first issue of its famous journal *Philosophical Transactions* appeared. Fellows of the Society were supposed to be elected, but the qualifying criteria were vague in the early days. Even though its founder members included men like Christopher Wren, Robert Hooke and Sir Robert Moray, the majority of its fellows were not professional scientists as such and at times it had the air and prejudices of a gentlemen's club.

But for all the bickering and claims that it sometimes wasted its time on the consideration of foolish and irrelevant projects, the Society had slowly established itself as one of the pre-eminent scientific bodies in Europe. By the mid-eighteenth century, its journal was edited by a specially appointed committee of qualified scientists and had published groundbreaking work by a breathtaking array of medical men, scientists, astronomers, botanists, chemists, mathematicians and natural philosphers; Isaac Newton, Edmond Halley and Benjamin Franklin were among the early contributors. Getting an issue or theory considered at one of its meetings or published in *Philosophical Transactions* was a significant achievement, something that many aspired to but comparatively few managed. For a relatively unknown country parson to try to gain access to this exclusive body was ambitious indeed.

But the Reverend Stone had two things going for him. The first was that his theories about willow bark were of genuine scientific interest, even though that his experiments were a little amateurish. The second was that he knew the Royal Society's current President personally, despite the yawning social gap between them. George Parker, the second Earl of Macclesfield, was an eminent mathematician and astronomer (he built one of England's finest observatories at his residence at Shirburn Castle in Oxfordshire, very close to Princes Risborough where Edward Stone had been born). He had been a Fellow of the Royal Society since 1722 and had served four terms on its council prior to his election to its highest office in 1752 at the age of fifty-three. He was also a Whig politician of some influence. His son, Lord Parker, was elected one of rural Oxfordshire's two MPs in a bitterly fought contest in 1754. Parker's agent at that election had been Edward Stone. What the Parker family thought of him can only be guessed at from a comment made by Sarah Parker, the Earl's daughter-in-law, who described him in a letter after the election as 'a good man, but slow for a celebrated canvasser'. However, there was a connection and Stone was determined to exploit it.

> I have no other motive for publishing this valuable specific, than it may have a fair and full trial in all its variety of circumstances and situations and that the world may reap the benefits accruing from it. For these purposes I have given this long and minute account of it, and which I would not have troubled your Lordship with, was I not fully persuaded of the wonderful efficacy of this Cortex Salignus in agues and intermitting cases, and I did not think that this persuasion was sufficiently supported by the manifold experience, which I have had of it.
> I am, my Lord,
>
> With the profoundest submission and respect.
> Your Lordship's most obedient and humble servant.
>
> Edward Stone.
> Chipping-Norton,
> Oxfordshire. April 25, 1763

As he signed the letter, sprinkled sand over the wet ink and called for one of his household to take it down to the London coach, Stone must have wondered again about its possible reception. Much would depend on His Lordship's reaction. If it got past him and was considered by the various eminent men on the Royal Society's governing council, then it might be chosen for a reading at one of the Society's famous meetings. He couldn't hope to be invited himself, of course, but there was at least a chance that it might subsequently be published.

Almost certainly, the Earl of Macclesfied did see it and passed it on, but he was unable to influence events much further. Early that summer he fell grievously ill and although he remained President of the Society until his death the following year, he took no further part in its transactions. It is to Stone's great credit, therefore, that his letter was considered on its scientific merits alone. But although it was read at a Society meeting on 2 June 1763, neither he nor his noble patron was present. A James Burrows Esq. sat in the chair. Not knowing Stone, he merely recorded the Society's thanks for 'a useful communication' and moved swiftly on to other business.

Later that year the letter did receive the further accolade of publication in the *Philosophical Transactions*. By no means an automatic honour, this was probably sufficient recognition for its modest author. But it must have hurt a little that the journal erroneously attributed it to a Rev. *Edmond* Stone (a mistake that has caused no end of confusion since) even though his signature at the end of the document was correctly transcribed.* However, he had his mind full of other things by then. That same year (1763) he wrote a short book called *The Whole Doctrine of Parallaxes*, which predicted where a transit of Venus could best be observed. More surprising was another letter he wrote to the Royal Society four years later, containing an ambitious algebraic resolution for cubic equations. But the Earl of Macclesfield was in his grave by then and the letter, dismissed as unoriginal, was

* By a strange quirk, a mathematician called Edmond Stone was an almost exact contemporary of Edward Stone. He was no relation but was considerably better known. Even today, some people persist in attributing the Reverend's discovery to his namesake. This confusion was finally resolved by William Pierpoint in 1996. See *Notes and Records of the Royal Society of London*, vol. 50, 1996.

never published. Stone died suddenly himself the following year, aged sixty-six.*

His first letter remained in the annals, though, and although it happened slowly he did make a few converts. In 1792, a Hertfordshire doctor called Samuel Jones reiterated the 'singular efficacy in the cure of agues' of willow bark, and by 1798 an English pharmacist named William White was reporting: 'Since the introduction of this bark into practise in the Bath City Infirmary and Dispensary, as a substitute for the Cinchona, not less than 20 pounds a year have been saved to the Charity.'

But the great irony of Stone's story is that although his rediscovery of willow's medical potential was a genuine milestone in the development of aspirin, he completely misinterpreted its effects. He believed he had found a cure for the ague at least as effective as quinine. But quinine attacks the plasmodium parasite that *causes* malaria (or at least it did until the parasite developed resistance to it), whereas what Stone had found was a therapeutic remedy for malaria's *symptoms*. Those symptoms – feverishness, high temperature, aching limbs and headaches – can be associated with a number of conditions, and the real significance of Stone's work was that he had stumbled on an extraordinary substance that could relieve them all. With hindsight, it is possible that some of his patients may not have had the ague at all.

None the less, years later, in the dazzling new chemical laboratories of Europe, his words would inspire a new generation of scientists. The work of harnessing his discovery could begin.

---

* Stone wrote four books in total: *The Reasonableness and Excellency of Abraham's Faith in offering up his Son, Oxford University Sermons 1732*; *The Whole Doctrine of Parallaxes Explained and Illustrated by an Arithmetical and Geometrical Construction of the Transits of Venus and Mercury over the Sun (Oxford and London 1763)*; *Remarks upon the History of the Life of Reginald Pole (1766)*; and *Discourses on some Important Subjects by the late Reverend Edward Stone*, published in 1771 by his son – also called the Reverend Edward Stone.

3

THE PUZZLE TAKES SHAPE

REVOLUTION, INDUSTRIALIZATION AND warfare – as dawn broke over the nineteenth century the world was being torn asunder by dramatic and violent change. It says much for human adaptability and endeavour that a century beginning amidst such ground-shaking confusion would end with the mass production of one of the most significant medicines ever discovered. There was nothing predestined about the industrial formulation of aspirin, any more than there was about the invention of the internal combustion engine or the building of the Suez Canal. But fortunately for us, the nineteenth century was a period when people frequently had the means, motive and determination to take an idea and turn it into reality. In the case of aspirin, this happened piecemeal – a series of minor, often unrelated advances, fertilized by the century's broader economic, medical and scientific developments, that led to one big final breakthrough.

Over the five decades that spanned the old and the new centuries, the world had been thrust into a bubbling stew of competing ideologies, political and social upheaval, unforeseen events and new ideas. The Enlightenment had brought scepticism and reason into fashion as the dominant philosophical tenets of the age. Capitalism, if not quite yet red in tooth and claw, had developed sufficiently to make aggressive competition, investment, entrepreneurship and the limited company the cutting tools of business. Utilitarianism, with its theory of the greatest happiness of the greatest number; the French and American revolutions, with their violent reactions to the *ancien régime*; the peculiar geological, political and economic circumstances that were bringing about the Industrial Revolution in Britain – all these things contributed to a world in which the old ways were being done away with, sometimes at a great cost in terms of human misery. Huge demographic shifts took place as people began moving from the countryside into the new urban centres. Warfare – always a great engine of change

– enveloped Europe as Napoleon sought to establish his supremacy over land and Britain fought back by reinforcing its dominance of the seas and securing its new colonies. And in between – often unnoticed at the time but just as significant in their way – were the smaller things, the gradual transformation of everyday life by technological innovation, scientific enquiry and economic enfranchisement.

It was against this background of almost unremitting change that the next intricate episodes in aspirin's story would be written. A new spirit of sceptical enquiry was abroad, a thirst for knowledge that would see chemistry emancipated from alchemy and pharmacy from the apothecary. New laboratories and scientific institutes would challenge preconceived ideas, increasingly spurred on by national rivalry and commercial competition. New businesses would turn these discoveries into products and develop new ways of selling them. Industrial scientists, pressed hard by investors and entrepreneurs, joined the ranks of dedicated amateurs and lofty academics in chasing science's great prizes. Eventually thousands of new technologies would emerge from this process, but such was the bewildering pace of events in the nineteenth century, and from so many different directions did changes come, that these discoveries would often be due as much to happenstance as systematic investigation. This underlined one of science's great truths: that only rarely do scientific breakthroughs result from a single moment of genius. More often they follow from small steps taken by individuals, each contributing one piece to the final solution of the puzzle. In aspirin's case, there were still a great many pieces to find.

What is this substance actually made of? It was a question asked time and again by nineteenth-century scientists as they began to grapple with the chemistry of a range of materials that had hitherto been taken for granted. A particular feature of these early pharmacological investigations was the rigorous re-examination of the *materia medica*, the naturally derived medicines and herbal remedies that doctors had relied on for years. Chemists were keen to identify and isolate the active ingredient in these substances, partly out of pure scientific interest, but in some cases for sound medical and commercial reasons too. Isolation, it was felt, could improve the potency of the drugs, help regulate their dosage and in the long term perhaps even make it possible to reproduce them synthetically and at lower cost.

The intellectual justification for this new science had been set out in 1797 by Johann Christian Reil, a medical philosopher who ironically would go on to become famous as one of the founding fathers of German psychiatry. In a paper entitled 'An article on the principles for a future pharmacology'* he outlined what must be achieved:

> It must factually and scientifically expose the changes in the conflict of a drug and a living body in its very origins. The drug also undergoes changes but these only interest us in that they help to explain the change to the human body . . . Scientific pharmacology requires a complete knowledge of the nature of the drug in all its conditions, especially chemical . . . We do still not know the specific and general constituents and especially not the quantitative condition of many medicines. As long as there are still gaps, a scientific treatment of pharmacology in all its parts is not possible . . . the only way to render pharmacology more complete is therefore to perform experiments, record the results accurately and subsume isolated experiences under higher laws.

On a more prosaic level, the new science may also have had its roots in the Napoleonic wars and the resulting shortage of cinchona bark. Way back in 1763, the Reverend Edward Stone had tested out willow as a possible alternative to cinchona because the latter was so expensive and difficult to get hold of. Fifty years later, several attempts to transplant cinchona trees to Europe had failed, and the continent was still reliant upon supplies from Latin America. Following the French Revolution, the rise to power of the Emperor Napoleon and the resulting conflict between Britain and France, those supplies were disrupted by a fierce Royal Navy blockade of commercial traffic across the Atlantic and a French stranglehold on the movement of goods throughout continental Europe. Furthermore, Spain's ability to maintain the trade (it was still one of the principal importers) was hampered by the fact that for several years it was either allied to France and therefore subject to the Royal Navy blockade, or occupied by the French and so unable to keep in regular communication with its overseas possessions. But malaria still plagued much of Europe and the demand for its most popular therapy was undiminished. Even when cinchona did get

---

* *Beitrag zu den Prinzipien fur jede zukünftige Pharamakologie.*

through, it was in such small amounts as to render it unaffordable to most people. The virtue of making such limited supplies go further was clear, and isolating its active ingredient was the next obvious step. But although the will was there, the know-how was not. Cinchona clung on to its secrets for some time yet.

But the experimental science that sprang out of these ambitions – particularly in France at the beginning of the century – allowed chemists to make giant strides in understanding the action of other natural drugs and in breaking them down. One of the first to be cracked was opium. In 1804, two French pharmacists, Armand Seguin and C. L. Derosne, isolated a raw crystalline substance from opium, although they weren't sure what it was. A year later a German pharmacist called Friedrich Seturner showed it was an alkaline substance and named it morphium. In 1809, another French scientist, Louis-Nicolas Vauquelin, did the preparatory work towards isolating nicotine. But the lions of this new chemistry were Pierre-Joseph Pelletier and Joseph Caventou, two Paris-based pharmacists. Between 1818 and 1821, they had a string of remarkable successes, isolating strychnine, brucine, veratrine, caffeine and then finally quinine, the active substance in cinchona bark which had eluded scientists for so long. All of these refashioned potent drugs were alkaline and so naturally came to be known as the alkaloids.

It was inevitable that before long the alkaloid chemists would get around to thinking about willow bark. By the end of the eighteenth century willow was sometimes being used as a cheaper substitute for cinchona in England and the knowledge had spread to the continent. Whether any of these scientists dug out and read Edward Stone's forty-year-old letter detailing his simple experiments isn't known, but old copies of the *Philosophical Transactions* were kept in libraries across Europe and it is certainly possible that they were referred to. What *is* known is that willow began to attract attention again and the race to isolate its key ingredient was on. Indeed, it seems to have been an obsession with some chemists; as each faltering step forward was reported in the journals, so others would be goaded back into the laboratory to try to improve on the efforts of their competitors. The first to have a go were two Italian scientists, Brugnatelli and Fontana, in 1826, but they found it difficult to prove that the impure lumps they had produced were the real thing. The first substantial breakthrough was made two years later by Joseph Buchner, Professor of Pharmacy at Munich University, who refined willow down into bitter-tasting yellow crystals. He obtained only a tiny amount of the substance but he christened it salicin (after *salix*, the Latin name for willow). In 1829, a French chemist called Henri Leroux refined the extraction procedure and managed to obtain around 25 g of salicin crystals from around one kilo of willow bark. He was then trumped in turn by another Italian, called Raffaele Piria, who in 1838 produced a more potent acid from the crystals, which he dubbed salicylic acid.

This was all cutting-edge stuff, advances made by a small group of specialist scientists each of whom knew, at least in outline, what the others were up to. In such cases, it is easy to forget that scientific discoveries can also come out of nowhere, from individuals working in isolation, unaware that their own pet project 'X' might one day have major implications for someone else's pet project 'Y'. This happened in aspirin's case. Not long after Henri Leroux finished refining his extraction procedures for willow, a Swiss pharmacist began working on a completely different plant.

Johann Pagenstecher was already something of an anachronism in 1830, one of Europe's last traditional apothecary-pharmacists, a figure who straddled the old era of individual empirical investigation and the new age of objective scientific analysis and laboratory chemistry. He lived on the edge of Switzerland's Bernese Oberland, where he ran a modest business administering cures and remedies to local people. It was a typical practice of its time and of many generations preceding it – for centuries, throughout Europe, apothecaries had been the equivalent of today's general practitioner, the closest thing to a local doctor for most people. But Pagenstecher was more than just a village quack. He was a man with a mission, devoting much of his life to a search for substances that would help relieve pain. Inevitably, he looked for these cures among the kind of folk remedies and herbal cures that formed part of his professional armoury. Like the Reverend Edward Stone seventy years earlier, he was a disciple of Paracelsus and the doctrine of signatures. The doctrine had been embellished over the years and now embraced the notion that each herbal remedy must contain at least one active ingredient to treat one specific disease. It was the same principle, though arrived at in a different way, which motivated many of the more sophisticated laboratory scientists working away in the continent's new scientific institutes.

One day Pagenstecher's attention fell on one of his favourite remedies, the meadowsweet flower, *Spiraea ulmaria*, which was thought to be of some benefit in treating toothache and rheumatism. If he could isolate meadowsweet's pain-relieving ingredient, it would mean an end to the laborious process of hunting for it in the fields around his home and make the remedy stronger and more widely available. So, working in a small room at the back of his premises, he began a simple distillation process, putting the shredded leaves through a vat of boiling water before transferring the juice to the simple glass bottles and test tubes that made up the rest of his apparatus. He kept at it for several weeks and eventually produced a colourless but sweet-smelling liquid – a tincture – which he believed contained the therapeutic essence of the flower. He wrote up a simple report, sent it off to a Swiss journal, and from time to time dispensed the distilled remedy to his local patients. And there the matter might have rested but for the fact that three years later his article came to the attention of one of the new alkaloid specialists, Karl Jacob Lowig of Berlin.

The paper fascinated Lowig because he was always on the lookout for new substances to play with. He managed to obtain some of the Pagenstecher tincture (which the Swiss pharmacist had named as an aldehyde) and set to work in his laboratory. After a great deal of experimentation he found that by adding oxygen to the aldehyde he was able to isolate an acid. This he tried out on himself and on volunteers (animal testing did not become fashionable until the end of the nineteenth century) and discovered that his new substance had some remarkable properties – it could reduce fever and alleviate pain. Believing he had identified a potent new remedy, he christened it *spirsaure* (after meadowsweet's Latin genus) wrote up his findings and waited for the plaudits to come in.* It was only later, when Raffaele Piria's work was published, that Lowig realized to his surprise that he hadn't found a new substance at all. He'd actually discovered salicylic acid, the same chemical that other scientists had been labouring to release from willow bark for many years. But his unforeseen results had at least made one thing abundantly clear. Whatever this substance was called and from wherever it was derived, it had undeniable medical potential – and *that* knowledge

* The name 'spirsaure' would later be reflected in the name of the drug that emerged at the end of this lengthy process of discovery: aspirin.

would tempt scientists back to it again and again in years to come.

Of course, Lowig's discovery was not the only unexpected contribution to this complex puzzle. Around that time, events were unfolding elsewhere that would have an even more dramatic effect on the development of aspirin and indeed of the whole pharmaceutical industry.

The Industrial Revolution that began in Britain in the latter part of the eighteenth century was founded on many things – a fortuitous mix of liberal politics, laissez-faire economics, the right geography and geology, innovative technologies and entrepreneurial business – all of which came together in the same place at the same time. But other more mundane matters played a part too, one of the most significant of which was Britain's ready supply of coal. Coal had long been used to heat people's homes – or at least the homes of those able to afford something produced in relatively modest amounts from small drift mines. Now, though, industrial quantities of it were required. Coal provided the energy that drove the new steam engines that were transforming the manufacturing process. Its derivative, coke, was used to run the blast furnaces that produced the iron from which those steam engines were made. As deep mines were sunk to meet this growing demand, so it became cheaper and more ubiquitous, and more ingenious uses were found for it. One of these came in 1792, when a Scottish engineer, William Murdoch, discovered that burning coal in a vacuum produced a bright flammable gas.*

---

* One of the most remarkable figures of the Industrial Revolution, William Murdoch was a Scottish engineer, who in his mid-twenties walked 300 miles from his home in Ayrshire to seek work at the Birmingham steam engine factory of James Watt and Matthew Boulton. Watt was not there at the time but apparently Boulton was so impressed by Murdoch's metal top hat, which he had made himself on a lathe, that he gave him a job. In 1779, the company sent him off to manage one of its engines at a tin mine near Redruth in Cornwall. Over the next twenty years Murdoch's home and workshops there became a centre of astonishing innovation. One of his many prescient ideas was for a road-going steam car. The company told him (a little unwisely) that there was no future in moving engines, but this didn't stop him terrorizing the people of Redruth with his prototype. Gaslight was his most notable invention. One day, while relaxing by his fire, he put some coal dust into his pipe and laid it among the embers. As coal gas was formed and shot out of the mouthpiece, Murdock saw that it shone brightly. In 1792, he installed the world's first gaslights in his home and works. Within fifteen years gaslights were going up over London's Westminster Bridge.

For factory owners with tireless new machines that could work though the night, it offered a cheap way of lighting their vast premises. Town corporations followed suit, lighting dim and foggy streets so that workers could get to work safely and on time to meet the shift patterns of this demanding age. Gas companies began to lay networks of pipes to carry the new fuel, and their lamps sprang up on street corners and were installed in front parlours. Indeed, one of the most visible signs of the nineteenth century's spreading industrialization – in Britain and in the rest of the developed world – was that it took place under the pale luminescence of artificial light.

Unfortunately, light was not the only product of coal gas. It left behind a noxious residue; a nasty, foul-smelling gunge called coal-tar that was difficult to get rid of and had no apparent uses. But then, once again, the ingenuity of investigative scientists provided the answer. Coal-tar, they found, was full of interesting chemicals and promised to create whole new areas of industrial exploitation. One of those scientists was a brilliant young German called Friedlieb Ferdinand Runge. At the age of twenty-five he had independently worked out how to isolate caffeine – though he was just beaten to it by Caventou and Pelletier in Paris. However, in 1834 he made his three most remarkable discoveries. The first was isolating a remarkable new substance from coal-tar, called aniline. The second was deriving a chemical from that aniline, called phenol. Phenol became better known as carbolic acid and was found to have antiseptic properties. Initially used in the sewage industry, carbolic later became famous when the Edinburgh surgeon Joseph Lister discovered its benefits in preventing post-operative infection. Tens of thousands of surgical patients in the nineteenth century who would otherwise have succumbed to gangrene probably owed their lives to Runge's discovery, and it remained in use until better ways of preventing infection were found.

But it was the third of Runge's discoveries that had the most marked effect. Phenol was not the only derivative of aniline that he found; he also worked out that he could get dye from it – the first organic dye, which he called aniline black. At the time this was mostly of academic interest; as a pure scientist, Runge had neither the means nor the motive to exploit this discovery himself. Others could, however, and the story of how that happened – as is so often the case in this complex narrative – is tied up yet again with the search for an alternative to quinine.

William Henry Perkin was born in London on 12 March 1838, and right from an early age it was clear that he had an interest in the way things worked. As a boy he tinkered with mechanics, taught himself to paint, and even at one point tried to build an engine – no mean feat in an age where such things were a rarity. His father, George, a builder by trade, hoped that these precocious talents would one day earn his son an elevated place in the same profession – as a draughtsman, perhaps, or as an architect. But it was not to be. A school friend showed young Perkin a few basic chemistry tricks with crystallizing substances and he was hooked. It was the start of an obsession that changed his life.

Around that time he was enrolled at the City of London School and attended classes on chemistry given by one Thomas Hall. The teacher recognized that Perkin had potential and suggested that he sign up as a student at the new Royal College of Chemistry. He was accepted at the tender age of fifteen.

The College was the consequence of a growing awareness that British science was lagging behind that of its continental competitors – Germany in particular. In the early 1840s, a public fund had been set up to establish a new institute to teach practical chemistry. Contributors included Gladstone, Disraeli and Albert, the Prince Consort. Through his contacts in his native land, Prince Albert persuaded a celebrated German scientist called Auguste Wilhelm von Hofmann (then only twenty-eight) to be the first professor at the Royal College of Chemistry. Hofmann was an inspirational teacher and a highly accomplished chemist in his own right, but he also clearly recognized talent when he saw it. At first Perkin was just one of his students, but by 1856 Hofmann had also made him his personal laboratory assistant at the College. Looking for challenges to occupy this precocious young mind, Hofmann suggested that Perkin try to synthesize quinine. The cinchona tree was still stubbornly defying attempts to grow it commercially outside Central America, and even though Caventou and Pelletier had isolated quinine some thirty-five years earlier, no one had yet been able to reproduce it chemically. What's more, this was a time of fresh colonial expansion for Britain, with the interior of Africa being opened up by a new wave of

Victorian explorers. Malaria continued to be a significant problem, and even the vast amounts of cinchona bark that were being expensively imported from Central America were insufficient to meet demand.

So, during a break from his studies and lab duties, Perkin took the problem home to the tiny laboratory in his digs overlooking London's East End docks. His first attempts were with a substance called allyl tolluidine that he attempted to oxidize with potassium dichromate. This failed, so he tried replacing the tolluidine with aniline, the now commonly available coal-tar derivate found by Runge some years earlier. This failed too, but by one of those flukes of science, the aniline Perkin used contained impurities. The resultant reaction left a black sludge that turned his test tubes a striking purple colour when they were washed in water. Perhaps he remembered Runge's own aniline dye derivative at this point, but something piqued his interest and he began to work out a way of extracting the wonderful new colour to see if it too had potential as a dye. We owe a great deal to that intuitive leap.

For almost as long as there have been clothes, people have been dyeing fabric to make it more attractive. For centuries, though, there was little variation in the colours people had available to them or the sources from which they were drawn. Until industrial chemistry came along, dye was produced from animal, vegetable or mineral bases, an expensive and time-consuming process. As a result the search for new dyes was always intense and the rarer colours highly prized. Purple, the colour of wealth, power and prestige, was a good example of this. Alexander the Great marvelled at the purple robes he found in the royal treasury in the Persian capital of Susa, when he took the city in 331 BC. Roman emperors dressed themselves in Imperial Purple extracted from the murex mollusc and restricted its use to the royal line. Pope Paul II introduced Cardinal's Purple, taken from the kermes insect in 1464, with a similar embargo on its use. Other colours had been found over the years, of course; the blue woad of the *Picti* in Gaul and the Celts in Britain, the brilliant red of cochineal developed by the Dutch in the seventeenth century, the yellow of querticon bark found in American oak in the late eighteenth century, but like the purple of kings and cardinals their production was fraught with difficulties. Even the most common colours, the red of madder, the deep blue of indigo and the yellow of saffron, were extremely hard to obtain. Saffron dates back to the ancient Minoan civilization of 1900 BC and came from the saffron lily found on Crete, but the process of extracting its dye involved simmering the stamens of the flower for hours and then only resulted in tiny amounts. Indigo came from an Indian plant of the same name and it too would release its colour only after weeks of complicated fermentation. Madder was even harder to obtain, coming from the rare *rubia tinctorum* plant found only in Turkey and the British West Indies. To produce enough madder dye for one bolt of cloth could take over a month.

People had always experimented with mixing these dyes, of course, developing different variants of shade and ways to make them fast. But as the Industrial Revolution took hold of Europe and millions of yards of cotton fabric poured out of the new machine-driven textile mills of Lancashire and elsewhere, the demand for cheaper and more interesting alternatives to the limited range of traditional colours became intense.

This is why William Perkin's discovery could not have been better timed. His further experiments with his new chemical solution, including dyeing several pieces of silk a brilliant purple, convinced him that this was a commercially viable product. With the help of his father and some advice from the dye industry, he obtained a patent and in 1857 set up a factory adjacent to the Grand Union Canal in London to produce it. He called his new purple *mauveine* or mauve, and within a few short years it became one of the most fashionable and sought-after colours on the market. It truly took off when Empress Eugenie, wife of Napoleon III, took to wearing it because she thought it matched her eyes. Queen Victoria followed suit, wearing mauve dresses to her daughter's wedding and to other grand public occasions. The press and public took note, and by the time he was thirty-five, Perkin was a rich man. Purple, once the exclusive preserve of cardinals, kings and emperors, was now available to all.

However, his contribution to science was more than just as a producer of one fashionable new colour.* By showing the way to others he became the father of a whole new organic chemical industry. Soon new coal-tar-derived colours were appearing throughout Europe, and out of the chemical processes they gave rise to, a whole range of innovative

* He went on to discover other aniline colours and also to identify a set of chemicals that would later form the basis of the perfume industry.

products would one day emerge – from explosives, flavourings and perfumes to plastics, paints and preservatives. Most significantly for our story, the pharmaceuticals industry that sprang up in the years following Perkin's discovery was a direct result of his pioneering example. He may not have discovered a way to synthesize quinine but indirectly he played a significant part in the development of aspirin. William Perkin didn't know it at the time, but he had placed another piece in the jigsaw puzzle.

Back in the world of pure science, there had been few advances since Buchner, Leroux, Piria, and Lowig had begun to unlock the therapeutic secrets of willow bark, meadowsweet and the other salicylic acid-bearing plants. Although their discoveries had received attention within a small peer group of scientists, they hadn't done much to change the public status of the substance. A few more pharmacists experimented with salicylic acid – for a time it was thought to be a possible food preservative or useful for keeping water fresh on long ocean voyages. A few more doctors tried prescribing it, chiefly for rheumatism and to reduce fever, although many still preferred to use salicin (the basic crystallized willow bark extract) instead. But for nigh on twenty years it had remained only one of a large number of compounds drawn from organic materials, none of which yet stood out as being uniquely beneficial. Its chief problem was that it was very unpleasant to take, irritating the mouth, gullet and stomach if swallowed in large doses. When it was isolated as an acid these side effects were more noticeable than they had been when it was subsumed in willow bark, and many people, having tried it once, were reluctant to take it again.

In 1853, however, around the time that young William Perkin was attending his first chemistry lectures in London, a French scientist called Charles Gerhardt came tantalizingly close to remedying this problem. Indeed, if things had turned out differently, we might have had aspirin forty-six years earlier than we did. A native of Strasbourg, Gerhardt was Professor of Chemistry at Montpellier University. Aged thirty-seven, he already had a brilliant reputation among his contemporaries in Europe, achieving a modest sort of fame in 1852 with a book called *Précis de Chimie Organique* in which he described his work with acid anhydrides, substances obtained when you separate acid molecules from water. His latest project was an ambitious attempt to classify organic compounds and to find out how they related to each other and what happened when he mixed them up (although with hindsight nineteenth-century chemistry sometimes appears to have been a logical and straightforward process, it was quite often just a question of chucking a few ingredients together to see what happened). One of the many substances Gerhardt examined under his microscope was salicylic acid. Aware of its uncertain therapeutic pedigree, he was curious to see how it was put together and to find out if tinkering with its constituent elements could enhance it in some way. The first thing he noticed was that the acid's molecular structure was made up of a central structure – a six-carbon benzene ring – with two attached components. These were a hydroxyl group of atoms (given the designation OH) and a carboxyl group of atoms (COOH). The hydroxyl set separated from the core when it rubbed up against the stomach wall and it was this that caused the dreadful gastric irritation that was the remedy's Achilles heel.

Gerhardt then attempted something very complicated. He tried to induce a reaction between a variant of salicylic acid, called sodium salicylate (chemically synthesized by a scientist called Henri Gerland a few months previously and a little easier to work with) and another substance called acetyl chloride. His aim was to replace the hydrogen atom of the hydroxyl group with an acetyl group – which in layman's terms was the equivalent of taking out a little bit of hydrogen and putting in a bit of vinegar. If his experiment had been completely successful, he would have found a way to reduce much of the drug's stomach-churning acidity. But it was a tricky process, nineteenth-century laboratory techniques being that much less sophisticated than they are today, and he only managed to obtain a crude and impure version of the final substance. None the less, Gerhardt was the first person to synthesize chemically a recognizable form of acetylsalicylic acid.* When we swallow an aspirin today, that is what we are taking – a compound of acetylsalicylic acid, or ASA as it is known. Regrettably, Gerhardt found the whole procedure so lengthy and tedious that he decided to shelve it.

Over the next few years others had a go either at refining the chemical

* An event marked by the citizens of Strasbourg if no one else. In 1956, on the centenary of Gerhardt's death, they raised a memorial to him in his home city.

synthesis of ASA or at coming up with a better process for replicating the basic salicylic acid. The former group did not get very far at first (the next best stab at it coming from a German scientist called Karl Johann Kraut some years later) but the latter had more success. In fact the process developed by one German scientist, Hermann Kolbe of Marburg University (who in 1859 managed to synthesize salicylic acid from sodium phenolate and carbon dioxide) proved so effective that it would later be taken up by one of his former students, Friedrich Von Heyden. He established a large factory, the Heyden Chemical Company, to produce it.

The jigsaw puzzle was coming together. But before the final picture would be revealed two more pieces were needed. Someone had to invest time, money, scientific expertise and industrial acumen into obtaining a commercially viable, side-effect-free formulation for the substance. And someone had to provide the medical impetus to convince physicians that all this frenzied activity was relevant to their patients, that this much-investigated family of chemicals (salicin, salicylic acid, spirsaure and acetylsalicylic acid) was as therapeutically beneficial as its supporters said it was. It is to a well-built Scottish doctor in top hat and frock coat that we turn next.

When the journalist and author Daniel Defoe made his famous tour through Britain in 1725 he found Dundee to be 'one of the best trading towns in Scotland . . . populous, full of stately homes and large handsome streets'. It is unlikely he would have recognized it 150 years later. A city that had once been famous for the manufacture of fine linen had given its heart and soul to more aggressive industrial processes. Jute was to blame. From the 1830s onwards, the importation of this raw plant fibre from India and its production into the strong fabric used in sacking, carpets and luggage, became Dundee's most important business.* Vast mills, their chimneys belching out steam and smoke, soon domi-

* There were others of course – engineering, textiles, shipbuilding etc. A popular epigram about Dundee later described it as a city of 'jute, jam and journalism – reflecting the importance of Dundee's Keiller company's marmalade production and the D. C. Thompson comic business (Dennis the Menace's Beanotown is set on the outskirts of Dundee). But for much of the mid-nineteenth century jute was king. Even the city's large whaling industry was partly dependent upon it, blubber oil being important for the treatment of jute fibre.

nated the landscape, and a huge migration of workers, many from as far away as Ireland, came to service them. Juteopolis was born.

Mill owners encouraged this massive inrush of people for obvious commercial reasons, and in their blinkered terms it worked all too well. Before long, the labour market was dramatically overstocked and wages in Dundee were among the lowest anywhere in Britain. Many of the workers were women and children and were exploited dreadfully; young men were frequently thrown out of their jobs before they could reach the age of eighteen and start demanding adult rates. And predictably, this new, vast, impoverished workforce completely overwhelmed the limited housing supply. With everything happening so fast, there was little time, money or inclination to build suitable accommodation for everyone. Short-term lodgings became long-term housing, tenements were divided and then divided again, and many of the once stately homes noted by Defoe became squalid warrens where hundreds of people lived.

The consequences were inevitable and tragic. By the mid-nineteenth century infant mortality in Dundee was among the highest in Scotland. Three big cholera epidemics hit the city between 1832 and 1854. Typhus and tuberculosis were endemic. Polio and rickets were commonplace. Eventually the city authorities awoke to their responsibilities and began trying to tackle the worst of the deprivation, but the problems of industrial poverty took much longer to repair than they had to create. The dank and overcrowded ghettoes at Overgate and Blackness would shame the city for some years to come.

Such was the situation that an ambitious young doctor called Thomas John Maclagan found when he arrived to take up a post as resident medical superintendent at the Dundee Royal Infirmary in late 1864. The city was in the grip of one of its periodic typhus epidemics at the time and the hospital must have been desperately pleased to have him on board. But it was something of a shock to Dr Maclagan whose medical experience up until then had been gained in more salubrious climes.

Although his parents lived only forty miles away, he was a well-travelled young man. His father, also a doctor, had married into a wealthy plantation family in Jamaica in his youth and on the proceeds had returned to Scotland to build up a large country practice at Scone in nearby Perthshire. This comfortable income had funded young

Maclagan's impressive education: an expensive private school in Perth, Glasgow University to read humanities, and then to Edinburgh for training as a doctor. He qualified as an MD in 1860 and set off around Europe for postgraduate study at medical schools in Paris, Munich and Vienna. This had left him well acquainted with the French and German languages as well as the latest advances in medical science, and with big ambitions. When a job opportunity arose in Dundee in February 1864, he decided to return to Scotland. The typhus epidemic was his first serious challenge – the medical superintendent's post carried certain public health responsibilities – and his recently acquired knowledge of European quarantine techniques was in demand from the outset.

Maclagan was appalled by the conditions he found in Dundee. Writing later, he described the 'filth and gloom' of some of the homes he visited. Some of these problems, he said, were not even the problems of deprivation but of the sheer inadequacy of the housing stock. 'On many occasions when visiting the typhus haunts I have found a family composed of father, mother and several children of all ages occupying the one room which formed their entire occupation and this not because of poverty but because they could get no more suitable place of abode.'

Amidst such squalor he was lucky not to become seriously infected himself. During the course of the 1865–6 epidemic, twenty-three of Dundee's doctors and nurses died of typhus. Indeed, on a later occasion while Maclagan was battling an outbreak of typhoid (a version of typhus that gives its victims acute dysentery and diarrhoea), the sewage pipes at the Royal Infirmary burst, flooding the hospital with contaminated water. He was fortunate that he caught only a mild form of the disease; his predecessor had died of the very same complaint in 1863.

By enforcing a basic and rigorous public health policy – quarantining patients and their relatives, destroying contaminated clothing and bedding, and so on – Maclagan and his colleagues gradually brought the typhus epidemic under control. And although there was little he could do about the widespread poverty in the city, he managed to improve sanitary conditions at the hospital to such an extent that it soon had the lowest patient-mortality rates in Scotland.

But, not surprisingly, when his contract expired in 1866 he decided not to seek another term. Instead he moved into general practice, bought a large house in Dundee's Nethergate Street and with his wife Isabella settled down into the life of a popular local physician. His experiences had affected him hugely, though, and over the following years he was to turn his considerable intellect on to the whole problem of fever and infection. Thus it was that in 1874 he began investigating one of the most prevalent and distressing complaints in the city at that time – rheumatic fever.

As we now know, this condition, with its painful arthritis-like symptoms, is caused by a strep-type infection. It was therefore bound to be rife in a place like dank and overcrowded Dundee where contagion was all too easy. But in the nineteenth century there were still a great many competing theories as to what caused rheumatic fever. Some physicians claimed that it was a result of too much lactic acid in the blood, others that neurosis was to blame. Maclagan's own strange pet belief was that rheumatic fever must be triggered by a parasite living in the muscles and fibrous tissues of the joints and the heart.* However, it wasn't the cause that occupied him so much as finding effective ways to treat it. In pursuing this interest he made one of his most valuable contributions to medical science. He began to experiment with salicin.

Until then, the extensive investigations into the chemical make-up of salicin and the search for an artificial synthesis of salicylic acid had taken place in the closed world of the laboratory and little of this knowledge had yet been exploited. The pharmaceutical industry was then still in its infancy and the close relationship that would later develop between science, medicine and the drugs business was some way in the future. Indeed, there were not yet any chemically produced drugs on the market, although their appearance was approaching fast. Nowadays, the process for a promising new drug is for it to be subjected to extensive medical trials before being put on sale. In 1874, even the idea of a trial was a novel one. The significance of Maclagan's experi-

* Right idea, wrong illness. Had Maclagan applied this theory to malaria – or the ague, to give its traditional English name – he could have cracked one of the great mysteries of medicine earlier than it was. The ague had often been confused with rheumatic fever in years gone by because of the superficial similarity of some of the symptoms. In 1880, some time after Maclagan came to believe that rheumatic fever was caused by a parasite, the French physician Alphonse Laveran discovered a parasite living in the white blood cells of a malaria victim.

ments was that they tested for the first time – in a scientific framework – the claims made by those who believed in the therapeutic benefits of the salicylates, including their use as an anti-rheumatic treatment. This is not to devalue Edward Stone's earlier, more modest research. He was after all the first person in modern times, albeit obliquely, to identify the potential value of the drug. However, Stone was a parson, not a doctor, and his work lacked the legitimacy of a full medical analysis. Likewise with the pharmacists and chemists who followed him. They had all contributed to the bank of knowledge about the substance, but their focus had primarily been on working around the chemical formulation of its key ingredient. It would take Thomas Maclagan to prove that it worked.

He had two sources of rheumatic fever victims: his own private patients at his practice in Nethergate Street and those who had been admitted to the Dundee Royal Infirmary. The DRI was an impressive institution for its time. It was built between 1853 and 1855 (at a cost of £14,500 in public subscriptions) in a grand Gothic style and stood on high open ground on the south side of the city. But there was more to it than just imposing architecture. It had 235 beds and during Maclagan's time as medical superintendent had become one of the first hospitals in Scotland to have separate medical, surgical and fever wards. Given the susceptibility of poorer Dundonians to rheumatic fever, many of them were to be found languishing on those wards or pressing to be admitted. Although he was no longer medical superintendent and strictly speaking had no direct access to its patients, Maclagan had become a governor at the hospital and so was not short of human guinea pigs.

In a curious echo of Edward Stone's experiments, Maclagan began his trial with the fabled doctrine of signatures firmly in mind; the creed of Paracelsus still held sway among many nineteenth-century physicians.

> Nature seeming to produce the remedy under climatic conditions similar to those which give rise to the disease . . . among the Salicaceae . . . I determined to search for a remedy for acute rheumatism [rheumatic fever]. The bark of many species of willow contains a bitter principle called salicin. This principle was exactly what I wanted.

He seems to have chosen salicin for his experiments in preference to salicylic acid because the less refined substance was known to be more palatable to the stomach. However, he did try two types of salicin, one derived from willow and one from meadowsweet.

> I had at that time under my care a well-marked case of the disease which was being treated by alkalies but was not improving.* I determined to give him salicin; but before doing so, took myself first five, then ten, then thirty grains without experiencing the least inconvenience or discomfort. I gave the patient referred to twelve grains every three hours. The results exceeded my most sanguine expectations.

The patient was forty-eight-year-old 'William R', who for the preceding four days had had a temperature of 103 degrees. In the days after taking salicin this came down rapidly and he was subsequently discharged from care. It was the first of many such recoveries that Maclagan witnessed over the following two years. The trials were rigorous – some patients were given the drug and others not, so establishing a simple control process that validated his results – and he was meticulous in insisting on regular doses until he was satisfied that the patient's pain had subsided and his temperature had become normal. This must have been a sore trial to the Dundee Royal Infirmary's bursar. In choosing salicin as opposed to salicylic acid, Maclagan was going for the more expensive option – at two shillings an ounce, salicin was nearly twice the price of the chemically produced alternative (one of the reasons why chemists had been keen to produce it in the first place). While some of Maclagan's wealthier private patients would have been able to bear the cost, many of the DRI's patients were there on charity and the hospital would have had to pick up the bill. Still, as a friend and colleague said of him some years later, 'he holds the strictest views as to his duty and carries out those views with a rigid and unflinching exactness',† and it is hard to imagine that he was ever bested in any arguments about money with his colleagues on the hospital's board of management.

Maclagan wrote up his results and sent them off to the *Lancet*, which

---

* Possibly quinine – sometimes used to treat rheumatics because of its antipyretic (fever-lowering) properties. Other more ineffective contemporary treatments included bleeding and infusions of mint or lemon water.

† A remark made by Sir Frederick Treeves, Maclagan's own doctor.

published his findings on 4 March 1876. Salicin, he declared, was a winner. He described it as 'quite apart from its antipyretic properties, the most effectual means yet for the cure of acute articular rheumatism and that it may even show itself to be a specific in the disease'. It had lessened the patients' symptoms – fever, inflammation and pain – and was undoubtedly a very useful addition to modern medical pharmacopoeia.

Maclagan's report had two immediate effects. The first was that the price of salicin began to shoot up – within a year it cost more than ten shillings an ounce. The second was that a flurry of other doctors began publishing their work on the same area. Almost contemporaneously, a German physician, Solomon Stricker, declared that *his* tests had shown that salicylic acid was also effective in treating rheumatism. Shortly afterwards, another German doctor, Ludwig Reiss, reported the same thing. The following year, Germain See, in Paris, claimed that the salicylates were useful not only against rheumatism but also against the chronic condition called rheumatoid arthritis. Furthermore, he added, they could be helpful in gout too. Others chimed in with claims that salicylic acid could help with headache, migraine and neuralgia. Between 1877 and 1881, four of the main London teaching hospitals began large-scale trials of the salicylates and subsequently introduced them as regular therapies. A more curious footnote came in the shape of a letter to the *Lancet* from one Dr Ensor of the Cape of Good Hope. Dr Maclagan might also like to know, he said, that the Hottentots of southern Africa had been taking willow bark to relieve rheumatic disease for years.

It is perhaps not so surprising that Maclagan's fellow professionals took such an interest in his findings. The *Lancet*, in which he published his results, was already established as one of the world's leading medical journals and its letters and articles were taken very seriously. Moreover, it should also be remembered that aside from quinine, the opiates and, to a different degree, digitalis (the heart drug which was extracted from the foxglove flower), there were still few *proven* treatments that physicians could rely on. Many of the other nostrums they dispensed had been around for hundreds of years and were often no more effective than a winning bedside manner. But salicin and the other salicylates were available, albeit in limited quantities, and now had the added endorsement of medical analysis. That gave them extra kudos.

Three years later, buoyed by his success and perhaps also keen to get away from the enduring grimness of his surroundings, Maclagan moved south to London. Although he retained an interest in the rheumatic diseases, fever and germ theory (hanging on firmly to his belief that rheumatism was caused by a parasite), from then on he turned to fashionable medicine. He opened up a practice in Cadogan Place and made his name as a society doctor, counting the Duchess of Albany and Thomas Carlyle among his patients and even becoming physician-in-ordinary to the Prince and Princess Christian of Schleswig Holstein. He died in 1903 of stomach cancer and was buried in a modest grave in Woking cemetery on the outskirts of London. Amid the many tributes he received on his death was this recognition from the *Lancet*:

> Decoction of willow bark had of course been known as a 'cure' of rheumatism for many years prior to 1876 and the use of salicin is nowadays almost superseded by the salicylates, but still Dr Maclagan would seem to have been the first to draw the attention of the medical profession to the use of salicin in modern days.

That is understating his contribution to no mean degree. By taking the salicylates seriously Maclagan helped create the climate in which they would finally be developed into one of the most significant medicines of the modern era. The thorny problem of mitigating the dreadful side effects still had to be solved of course, but fortunately the foundations for dealing with that had already been laid by Messrs Runge, Murdoch and Perkin, some years before. The last piece in aspirin's jigsaw puzzle was about to be placed by – of all things – the German coal-dye industry.

4

THE BIRTH OF A WONDER DRUG

AROUND ABOUT THE time that the Luxor adventurer Edwin Smith was gloating quietly over his newly purchased but very ancient medical papyri, artefacts of a more modern age were drawing the crowds several thousand miles to the north. Housed in a hall in leafy South Kensington, the London International Exhibition of 1862 was a showcase for the very latest products of the mid-Industrial Revolution.* There were thousands of remarkable and interesting things to see – steam-driven water pumps, cut-crystal bowls and fine porcelain, photographs, safety matches, microscopes, toy soldiers, ingenious mechanical devices of every possible shape and size. These were all hugely enjoyed by the visitors, most of whom were the same inquisitive Victorian middle classes who always attended events such as this: for a few shillings they got a window on the world of scientific and technological achievement and enjoyed an improving day out in the process. But there were others with more professional interests; retailers looking for new products to buy and sell, manufacturers sizing up the opposition and, of course, the press, there to serve the huge public demand for information about what was *new*. As they flicked though their catalogues many would have made a mental note to visit one particular presentation on the ground floor. There, amidst an array of ribbons, shawls, hats and a large purple pillar of dye, William Perkin was showing off his new colour mauve to the world. It was an eye-catching display and probably would have attracted attention in any case, but interest had recently been given a huge boost by Queen

* Not to be confused with the celebrated Great Exhibition of 1851, which occupied the spectacular Crystal Palace in Hyde Park. The International Exhibition built on the successes of the earlier event and like those that followed became the precursor of the modern Expo movement.

Victoria's decision to attend the event wearing a vivid mauve gown – the kind of celebrity endorsement that today's manufacturers can only dream of. Perkin's dye was a *cause célèbre*.

His admirers were not restricted to Her Majesty's subjects. The exhibition had attracted sizeable contingents from France, Italy, the Low Countries, the United States and Germany, many of whom made a point of stopping by to chat with the inventor of mauve. Flattering though it must have been to be the subject of so much international attention, it's possible that Perkin later remembered those visitors with a rueful smile – particularly the Germans. Because if there was one country that he should have had cause to be wary of, one country with the means, motive and opportunity to take advantage of his ideas, it was Germany.

Germany was in the throes of becoming a unified nation. In 1834, most of its thirty-eight independent states had bonded together in one customs union, and ever since then, under the austere eye of Otto von Bismarck, they had been discovering their common destiny. Although that process would not be finally completed until 1871, to all intents and purposes Germany was already a single nation, and like many new nations was hungry to make its mark on the world. Its industrial development had been slower than Britain's – hampered by the Byzantine commercial laws and treaties that governed earlier relationships between its separate states – but it had begun to catch up fast. It also had one particular advantage: its scientists were the best trained in Europe. Universities and institutes at Leiden, Marburg, Berlin, Munich, Heidelberg, Göttingen, Freiburg, Dorpat, Kiel and elsewhere, had all made science, and in particular chemistry, an important part of the curriculum. The fact that the first professor of Britain's Royal College of Chemistry, Auguste Wilhelm von Hofmann, was a German was an acknowledgement of that supremacy, as indeed was the fact that in 1864 Hofmann returned to his native land to take up a professorial chair in Berlin. Almost all the top chemists of the day were trained in Germany or were taught by someone who had been.

Being able to call on this expertise gave German business a unique advantage when it came to exploiting new technologies, and resulted in whole new industries and commercial opportunities. The synthetic dye industry was one of them. When news of Perkin's discovery spread through Europe, it was a French chemist called Verguin who reacted first, producing fuchsine, or magenta as it came to be known. But it was in Germany that Perkin's ideas had most impact. The German textile industry had long resented British dominance of the production of natural dyes and the high prices they were forced to pay and had been seeking alternatives. Now, with abundant coal being produced in the Ruhr and the scientific wherewithal to develop the new aniline chemistry, German businessmen saw from Perkin's example how this demand could be met. They grasped the opportunity with both hands. Coal-dye companies sprang up all over Germany, a host of new synthetic colours were rapidly discovered, and before too long German dye companies were masters of the industry.

Friedrich Bayer and Johann Friedrich Weskott set up one of the first of those companies. Bayer came from a family of silk weavers in Barmen, about twenty-five miles from Cologne. Born in 1825, he was the only boy among six children and it was fairly obvious that whatever he did in life would have some connection with textiles. His first job was as a chemical merchant's apprentice, but by the time he was twenty-three he had begun his own business trading in natural dyes. By 1860 the firm was flourishing, with trade links throughout Germany and Europe, and Bayer was looking to expand.

Johann Friedrich Weskott also came from a textiles background. His family had originally moved to Barmen because the Wupper river offered good water supplies for their bleaching business. Like Bayer, he was an ambitious young man and had opened a cotton-yarn dye works in 1849. In 1863, both men realized the huge potential of the new synthetic dyestuffs promised by Perkin's discovery and agreed to combine their expertise in one joint venture. Friedrich Bayer & Company was born.

Their first experiments were not a great commercial success. The partners managed to produce their own version of fuchsine dye in small premises attached to Bayer's house but lost much of their initial profit in compensation payments when irate neighbours complained about waste chemicals polluting the local drinking water. They shuttled around between various temporary homes after that, finally settling into more appropriate accommodation on the banks of the Wupper river. There they learnt to dispose of their waste more sensibly (or at least more discreetly) and began to branch out. The business grew steadily rather than spectacularly for the next two decades, producing a number of new dyes out of the scientific advances of others rather than their own research. Aniline blue was one early success, alizarin (an orange-red) another, but competition in this new industry proved tough and margins gradually tightened. In 1880 Bayer died, followed a year later by Weskott, and the reins were picked up by Carl Rumpff, Bayer's son-in-law. Rumpff had worked for a while in the United States, establishing his own small coal-dye business, but threw in his lot with the German firm when he married into the family. Now, seeing that the business was struggling, he realized he had to hire some new blood to get the company moving again. His first step was to raise capital by selling its stock to the public, in the process renaming it Farbenfabriken vormals Friedrich Bayer & Company (the Dye Factory formally known as Friedrich Bayer and Company). Then he went looking for scientists. His plan was to sponsor a few young chemistry graduates through their doctoral or post-doctoral studies, the *quid pro quo* being that the scientists would spend a year conducting research into new dye combinations on behalf of the company. It was a risky but, as it turned out, inspired move. One of these chemists was Carl Duisberg.

In his lifetime he would oversee the creation of the most powerful chemical and pharmaceutical combine in history – an industrial empire that would dramatically influence world events, employ hundreds of thousand of people and, if that were not enough, finally get round to mass-producing a commercial product out of something that people had been tinkering with for 6,000 years. That product was aspirin.

Duisberg was born on 29 September 1861 in a neat little house on Heckinghauser Street in Barmen. His father, a simple man with conservative ideas, had a small ribbon-making business with a couple of looms, and with the help of his wife Wilhelmina supplemented the family income with a small dairy. For his part, young Carl did his chores, dutifully attended the small local schools and generally kept his head down – until he discovered chemistry. He began to take an interest in his future profession at secondary school, probably after attending his first science lesson at the age of fourteen. Unfortunately, his father had other plans. As far as Duisberg Senior was concerned, his son had a duty to follow in the family firm and not waste his time with any expensive and nonsensical notions about science. But it's often said that behind any successful man there is a determined mother, and now (and not for the last time) Wilhelmina came to Carl's rescue. After many furious rows, it was she who persuaded her husband to let their son carry on with his education and that the financial sacrifice would be worth it in the end.

From then on Carl was in a hurry. He knew what he wanted to do, but not how long his father would let him pursue his ambition. He took his high school diploma at the age of sixteen (a year earlier than his contemporaries), rushed through a chemistry course at Elberfeld technical college and then enrolled at Göttingen University. He lasted a year. Into those twelve months he packed as much study as another student would do in three. He took every course going and managed to complete his thesis in record time. It was only then that he found out that he was ineligible for a degree because he didn't have the requisite Latin to pass a mandatory test. In a fury he transferred to another college at Jena, coming under the wing of Anthon Geuther, a leading academic chemist of the day, who insisted that his supercharged student slow down a little and learn the basic laboratory techniques that he would need in later life. Although he was fretting to get on, he took this extra delay in his stride and won his doctorate on 14 June 1882, at the age of twenty. He and his fellow students celebrated so boisterously that the police were called and the future mogul was fined ten marks for breaching the peace.

Duisberg could ill afford even that. He was now a qualified chemist, but until he could find a job he would still be financially dependent on his father, who never missed an opportunity to point out what a disastrous career choice his son was making. So over the next few weeks he scoured the job advertisements in the academic and trade press and sent hopeful application letters to every company and institute he could think of. Unfortunately, there were more chemistry graduates coming out of the universities than there were posts at that time, and he got nowhere. In desperation he accepted a poorly paid temporary job as assistant to Anton Geuther, and then, thinking that his lack of military service might be proving a hindrance to his applications, he signed up as a one-year volunteer with the First Bavarian Regiment. But twelve months later he was back living uncomfortably under his father's roof, and still without a job.

Anyone less determined might have given up at that point and tried their hand at something else, but as he would so often show later in life, once Duisberg got an idea into his head it was very difficult to shift. Closing his ears to the protestations at home, he held firm, wrote more letters, and finally fortune came calling in the shape of Friedrich Bayer & Company. Its boss Carl Rumpff made the young chemist an offer: spend a year as a research fellow sponsored by the company at 150 marks a month, and if something came of this, then maybe, just maybe, there might be a job at the end of it.

It wasn't ideal; the money was pitiful and the company was based just a few miles up the road in Elberfeld – for obvious family reasons Duisberg had been hoping to get a job as far away from Barmen as possible. But with nothing else on the horizon he accepted, consoling himself that at least it was a foothold in the dye business where much exciting work could be done.

His first assignment was a seemingly impossible one, to try to find a chemical synthesis for indigo – something that dyestuff chemists had been trying for decades without success. But if it was meant as a test of his character, he passed. He got down to work with a will, and something about his determined, if doomed, search for this dyer's Holy Grail must have impressed his new boss. On 29 September 1884, his twenty-third birthday, Duisberg was able to tell his father that he had been offered a job with a handsome starting salary of 2,100 marks a year. This must have been all the sweeter as he had just started courting the girl he would eventually marry – Carl Rumpff's niece.

With his future thus assured, it was time for Duisberg to return the compliment. His next task was to reproduce Congo Red, a scarlet dye that was then proving very popular with the cotton industry. As it happened, Congo Red had been synthesized a year earlier by another Friedrich Bayer chemist who had promptly left the company, patented the dye himself and sold the rights to a competitor. However, a loophole in German patent law at that time allowed a company to produce someone else's product if they could come up with a slightly different way of making it. Obviously, the first thing any successful inventor did was to try to circumvent this possibility by patenting every other production process imaginable, but sometimes they would miss something and their competitors would pounce. As might be imagined, this all too frequently ended in the courts (the law was later amended) and so the ideal was to find a method that was as demonstrably unlike the original as possible. Duisberg managed to find a dye equivalent to Congo Red within a few weeks and, better still, used a process that was sufficiently different to satisfy the lawyers and hence make his new employers a great deal of money. When he repeated the trick with another dye the following year and then came up with a third dye formula a short while later (this time an original one of his own making), Rumpff and his fellow directors realized that Duisberg was a chemist of rare talent who needed support. He was put in charge of all the company's research and patenting and given a new team of chemists to work under him. Charged with his success, one of Duisberg's first tasks was to look for new areas into which the company could expand.

It was then that he heard of Antifebrine.

The small Rhineland town of Hoechst lay about sixty miles away from Barmen and was home to one of Bayer's rivals – a synthetic dye company run by two chemists, Eugen Lucius and Adolf Brunning. In 1884, a doctoral student called Ludwig Knorr approached the pair. He had been tinkering with aniline chemistry as part of his studies and in the process had stumbled on a substance that he thought had potential as an antipyretic (a fever-lowering drug). He took it to Lucius and Brunning because they had previously tried marketing their own aniline-based antipyretic tonic called Kairin. This was supposed to be a synthetic substitute for quinine, the naturally derived and very expensive product that every chemist in Europe had been trying to replicate for years. Unfortunately, as in most of those attempts, Kairin had caused very unpleasant side effects and had to be withdrawn from sale. However, Knorr's compound seemed to be more promising, so Lucius and Brunning bought the rights. They began marketing it as a substance they called antipyrine. The compound did reasonably well, and even though it too turned out to induce unpleasant gastric reactions in those who took it, its brief commercial success showed that this was a line of chemistry that could profitably be followed up.

Then one day in 1886, two Strasbourg hospital doctors, Arnold Cahn and Paul Hepp, sent off an order to the Kopp wholesale pharmacy. They were treating a patient who had intestinal worms and wanted to give him the standard treatment – a substance called naphthalene. But there was a mix-up in their order at the pharmacy and they were sent – without them knowing it – another chemical called acetanilide. This was an acetylation of aniline – a by-product of coal-tar distillation, commonly used in the dye industry. It was most definitely not a medicine and had never been given to humans before. None the less, that's exactly what Cahn and Hepp did, in the mistaken belief that it was naphthalene. It was only when it became clear that the substance was having no effects on the intestinal parasites that they sat up, took notice and discovered the mistake. But to their surprise and gratification, they also noticed that their patient's temperature had fallen quite noticeably. Clearly, acetanilide was causing this effect.

Paul Hepp's brother was a chemist at Kalle & Company. It was one of a number of firms that produced acetanilide and other chemicals for the coal-dye industry and the doctors asked whether it would be interested in marketing acetanilide as an antipyretic drug. After conducting their own tests and taking a look at the healthy market for other fever-reducers like salicylic acid and antipyrine, the company's executives said that they would. However, there was a problem. There was no secret about making acetanilide; all of their competitors produced it too. If they went ahead and produced a drug under that name, then so would their rivals and all commercial benefit would be lost. So instead they coined a new brand name for the product, Antifebrine, and promptly trademarked it as their own. It was a revolutionary move.

Until Antifebrine came along, the drugs sold by pharmacists were usually known by their complicated chemical names and were described as such in the medical literature that doctors read to inform themselves about new therapies. Even though most doctors didn't have a clue about the chemistry involved, they traditionally used these same terms on the prescriptions they filled out for their patients, leaving it to pharmacists to decide which supplier they obtained the substance from. However, when a drug appeared with a nice simple name like Antifebrine, doctors found it easier to remember than the generic term acetanilide and put it on their prescriptions instead – even though Antifebrine and acetanilide were exactly the same substance.

Pharmacists knew this of course, but they were powerless to do anything about it because a doctor's prescriptions were legally sacrosanct and had to be followed to the letter. As a result, and to their considerable fury, pharmacists soon found that they were having to supply copious amounts of Antifebrine from Kalle & Company, and being forced to ignore the identical compound acetanilide that could be had much more cheaply from other suppliers. The patients were paying through the nose, but the company was making huge profits.

For Carl Duisberg, casting around for a new area of business for Bayer to get into, the commercial success of Antifebrine was a revelation. Could they pull off the same trick? He then remembered that lying around in the back yard of their Elberfeld plant were 30,000 kilos of a substance called para-nitrophenol. This was another waste product of the dye industry, similar to acetanilide, and Duisberg wondered whether an equally successful antipyretic could be produced from it. He gave the task to Oskar Hinsberg, one of the other doctoral students brought in by Carl Rumpff. After a few weeks Hinsberg came back with some very promising results. He'd produced a substance he called acetophenetidine that showed all the signs of being a more powerful antipyretic than acetanilide and what's more seemed to have less harmful side effects.* After testing it on volunteers among the Elberfeld chemistry team and a brief clinical trial, the Bayer board agreed to put it into production. Mindful of the success of Antifebrine, Duisberg decided to give it a memorable brand name, Phenacetin.

Phenacetin was the infant pharmaceutical industry's first big success. Unlike the few other early drugs, which had come about as a result of scientific investigation by academic chemists or medical men, it was purely the product of in-house *industrial* development, invented and marketed with the intention of making money. Of course, it had to perform therapeutically – something it was fortuitously able to demonstrate a few months after its launch in February 1888, when a major influenza epidemic swept over Europe and North America and demand for fever-reducing treatments was particularly high. But it owed its existence to a commercial ethic rather than a scientific one. The origins of today's multi-billion-pound global pharmaceutical business can be traced back to that moment.

---

* Both drugs were reasonably effective fever-moderators but both had significant side effects. In large or continuous doses they caused quite serious kidney damage and turned the patient's skin an alarming blue colour. However, these side effects were much more marked in acetanilide (Antifebrine) than acetophenetidine (Phenacetin), which allowed Bayer to claim that its drug was better tolerated by patients.

The drug made Bayer a great deal of money over the next few years, even though the company was badly stretched to meet the demand. It was still, after all, a dye-making business. The first batches of Phenacetin powder were brewed in hundreds of discarded beer bottles in a shed at the back of the Elberfeld plant and had to be hand-poured into glass containers for distribution to pharmacists and hospitals. Duisberg, though, had seen the light. Later that year the company produced its second drug, a sedative called Sulfanol, its complicated generic chemical designation, diemethylmercaptodimethylmethane, underlining the commercial imperative of giving drugs a catchy brand name. This too was successful and led to the formulation of a third successful drug – a more advanced version of Sulfanol called Trional. It was clear that the pharmaceuticals business was the one to be in.

The death of his patron, Carl Rumpff, in 1890 gave Duisberg effective day-to-day control of Farbenfabriken vormals Friedrich Bayer, the rest of the board being happy to defer to someone so clearly sure of his touch. One of his first decisions was to set up a discrete pharmaceutical division and build a proper laboratory for its chemists. Until then the rapid expansion of the company into drug production and the concomitant increase in the number of research scientists had meant miserable overcrowding throughout the Elberfeld plant. Chemists had had to find space in bathrooms, corridors, under the stairs and even in an old wood shed. There was a shortage of technical equipment like retorts and pipettes and only three sinks in which to wash them. One chemist, Heinrich Volkmann, had to perform his experiments out in the yard, much to the alarm of the factory foreman in charge of fire safety. The new facility changed all that. Built at a cost of 1.5 million marks (an astonishing sum at that time), it had three storeys and several large rooms, each with cubicles for twelve chemists. The individual work spaces were equipped with reagent chemicals, water, gas and compressed air and had efficient ventilation and extraction – hitherto it hadn't been unknown for chemists to slump unconscious over their workbenches, overwhelmed by noxious fumes.

Meanwhile, Duisberg's personal life underwent a transformation too. He married Carl Rumpff's niece, Johanna, and moved into a sumptuous house in Elberfeld, which he began to fill with paintings and fine furniture. The first of his four children soon appeared, to whom he extended the kind of easy-going indulgence that his own father had so singularly failed to show him. Things were looking good.

And so the scene was set for the end of the first act of this drama. From the very first uses of medicinal willow bark through the work of an unknown Ancient Egyptian physician, to Hippocrates, the Reverend Edward Stone, the scientific search for the secrets of salicin and salicylic acid, the rheumatic experiments of a Dundee doctor, William Perkin, the birth of the coal-dye industry and the development of the world's first brand-named medicines – all roads had led to this place, at this moment in time where, finally, aspirin was about to be born.

Every scientist who joined Bayer's drug research laboratory was asked to read a document, signed by Carl Duisberg, setting out his responsibilities. Their task, he said, was to:

> Find new ways of presenting familiar, especially patented pharmaceuticals by making use of the whole range of chemical, pharmaceutical, physiological and medical literature and also to discover new, technically utilisable physiological properties in new or familiar substances, so that the dye works are in a position to include the specialities of competing firms in their manufacture and bring to the market and introduce new pharmaceutical preparations.

Perhaps conscious that new arrivals might be a little intimidated by this ambitious mission statement, and knowing too from his own experience the role that chance so often played in scientific discovery, Duisberg went on to soften his words with a little reassurance.

> Great practical results are not and cannot be expected from each individual gentleman. The technical effect depends more often than not on coincidences that no one can safely predict. Each individual is merely required to work inventively and create innovation.

It was just as well, then, that he had employed some of the most inventive and innovative minds around. By 1890 the laboratory had been organized into two sections – the pharmaceutical group, which came up with the ideas for new drugs, and the pharmacology group that tested them. The first head of pharmacology was Wilhelm Siebel, a former assistant to the famous bacteriologist Robert Koch who discovered the bacilli that cause tuberculosis and cholera. When Siebel was forced to retire, ironically because of tuberculosis, his place was taken briefly by Hermann Hildebrandt, a senior Bayer researcher, and then by Heinrich Dreser, an associate professor of pharmacology from Göttingen University. Dreser's opposite number in the pharmaceuticals section was Arthur Eichengrün, another former academic and later to be one of Bayer's most innovative employees with a host of patents to his name.

It is with Dreser, Eichengrün and a young chemist called Felix Hoffman, one of Eichengrün's colleagues in the pharmaceuticals division, that the development of aspirin is most closely associated. Their relationship, particularly the frosty one that developed between the two section heads, would one day be the basis of much acrimonious recrimination, but in the late 1890s they were the trio behind the most successful drug the world has ever known.

The youngest of them had been the first to join the company. Felix Hoffman was born into a comfortable petty bourgeois family in Ludwigsburg in 1868. Like so many of his contemporaries he was bedazzled as a child by Germany's pre-eminence in the chemical sciences – a source of great national pride at the time – and was determined to make it his field too. At the age of twenty he was sent to Munich University to study as a pharmaceutical chemist and stayed on to conduct postgraduate research. On 1 April 1894 he was hired by Farbenfabriken vormals Friedrich Bayer & Company.

What this mild-mannered twenty-six-year-old made of Duisberg's mission statement is not known, but he seems to have fitted in with little fuss. It helped that a pleasant collegiate atmosphere prevailed in the pharmaceutical section; the team worked to a principle known as '*Etablissementserfindung*' in which close communication and idea sharing between colleagues was seen as vitally important. Although deference to the head of the laboratory was taken for granted, there was little formality otherwise. Even the dress code was comparatively relaxed – most worked in lounge suits or shirtsleeves rather than white coats, sporting straw boaters as a kind of raffish symbol of their trade. The appointment of Arthur Eichengrün to run the section in 1896 did little to change this. A flamboyant and charismatic man, as well as a brilliant chemist, Eichengrün knew his colleagues needed intellectual space to be effective. Once he had assigned them a task he left them alone until they had something concrete to show him or needed his help and encouragement. It was in identifying those tasks in the first place that he showed his true genius.

Heinrich Dreser was the most formidable of the aspirin trio. The epitome of an eccentric German professor, he was as unworried by outward appearances as anyone else; his personal idiosyncrasy was a beloved but overweight dachshund that he sometimes dragged to work (where it would sit, wheezing heavily, under its master's workbench). But this one engaging soft spot aside, he wasn't an easy man to work with. He could be bitingly sarcastic at times, was often seen as both pedantic and wilful in his judgements, and was heartily disliked by many of his colleagues who saw him as an authoritarian loner. None the less, he commanded their professional respect for his absolute determination to make his pharmacology section effective and efficient. It had been created to ensure that any drug product leaving Bayer did so with as few harmful side effects as possible, and Dreser took this gatekeeper's role very seriously. He set up a strict regimen of clinical testing and animal experimentation (he was one of the first pharmacologists in the world to do so), introduced exacting bacteriological and toxicology procedures, and insisted that the often subjective judgements of his peers in the pharmaceutical section down the hall be subject to the sternest examinations he could devise. Clashes with his more flamboyant opposite number, Arthur Eichengrün, were therefore both inevitable and frequent.

The exact role that Hoffman, Eichengrün and Dreser played in developing aspirin has been the matter of much debate and revolves around three key questions: who first had the idea to develop the drug; how original was the research on which it was based; and what subsequently happened to it?

If a much promulgated company legend is to be believed, both the idea and the research stemmed from Felix Hoffman. According to this account, Hoffman began looking for a formula because his father suffered dreadfully from chronic rheumatism. He had been taking sodium salicylate to ease the pain but the remedy was playing havoc with his stomach and so for years he had been begging his chemist son to come up with a less corrosive alternative. Hoffman set to it with a will and came up with a totally original solution. Heinrich Dreser then carried out some tests, declared the results satisfactory, and aspirin was born.

It is a nice, neat anecdote, but unfortunately little of it is true.* This account only emerged several years after the event, when (as will become clear in a later chapter) it was almost certainly influenced by political and commercial expediency. The real story, based on all the available evidence (and much more interesting) goes as follows.

In 1897, shortly after he arrived at Bayer (and presumably taking Duisberg's strictures about finding new formulations for familiar substances to heart), Arthur Eichengrün decided to try to find a version of salicylic acid that was free of unpleasant side effects. As the leading scientist in the laboratory, it was he who assigned Felix Hoffman to the task.

It was a natural target for any ambitious pharmaceutical chemist to focus on. Following the clinical tests of Maclagan, Stricker and others some twenty-five years earlier, salicin, salicylic acid and particularly sodium salicylate had gone on to become widely used treatments for rheumatic fever and arthritis. But they were all still distressingly corrosive on the stomach and in some cases caused other side effects like tinnitus. Clearly, neutralizing these problems whilst maintaining the substance's therapeutic benefits could produce a drug with some added commercial value.

There was not, it must be said, anything too remarkable about the decision to initiate this investigation. Bayer's chemists were continually experimenting with hundreds of such substances and for all the sophistication of their new laboratory it's easy to forget how haphazard much of this research actually was by today's standards. The scientists had a basic kit of parts of a few hundred chemical groups and the accumulated wisdom from their own observations and from scientific and medical journals about the effects (real or imagined) these might have on the human body. Different combinations were tried out; something was added here, something reformulated there. Sometimes they came up with a promising lead; more often than not their trials ended in failure. With so many experiments in progress it wasn't always easy to be systematic, and quite often Eichengrün's team would be directed to

* The part about Hoffman's father suffering from rheumatism is probably accurate and he may even have benefited from his son's work, but that's about it.

work on something without even knowing what it was for. They would add the task to the pile until they had a chance to look at it and then, perhaps by luck or ingenuity, stumble on a useful product. Since the advent of Phenacetin and Sulfonal, a number of other successful drugs had emerged in this way, including Aristol, an antiseptic, and Somatose, a sleeping draught, but hundreds of others had been rejected.

It is highly likely then, that when the assignment landed on Felix Hoffman's desk, he followed routine practice for anyone embarking on a new piece of work and went and did some research in the library. There he would have quickly found an 1853 copy of *Annalen der Chemie und Pharmacie*. This was the journal in which Charles Gerhardt, Professor of Chemistry at Montpellier University, had reported his attempts to reduce salicylic acid's nasty gastric side effects by crudely synthesizing acetylsalicylic acid, or ASA. In later editions of the same journal Hoffman would have come across similar attempts by other scientists, including one reasonably successful effort by Karl Johann Kraut in 1869. From this formula, the Heyden drug company was already producing its own unbranded version of acetylsalicylic acid.

To what extent Hoffman would have been influenced by these articles can only be guessed at. In the end it doesn't matter much, because one thing is indisputable. He began to replicate the experiments, playing variations on the same chemical theme. Thus it was that he wrote himself into pharmaceutical folklore on 10 August 1897 when he jotted the following into his laboratory journal:

> When salicylic acid (100.0 parts) is heated with acetic anhydride (150.0 parts) for 3 hours under reflux, the salicylic acid is quantitatively acetylated. After distilling off the acetic acid one obtains the above in the form of needles, which, when crystallized from benzene, melt at 136 degrees (value in the literature is 118 degrees). In contrast [with] the literature reports, my acetyl product no longer gives a reaction with ferric chloride, which readily distinguished it from salicylic acid. By its physical properties, e.g. its sour taste without being corrosive, the acetylsalicylic acid differs favourably from salicylic acid, and is now being tested in this respect for its usefulness.

This is complicated for a non-scientist, but in layman's terms Hoffman was saying that he had found a way of making ASA that neutralized the chemical element of salicylic acid responsible for its stomach-churning acidity. In essence he had done what Gerhardt had done, only much more effectively.

So far so good, but now the new substance had to be handed over to Heinrich Dreser's pharmacology department for testing. Arthur Eichengrün was present several weeks later when the ASA preparation was first put through its paces and was delighted to see how effectively it performed. It was obvious to him that it should go on to the next stage – clinical trials. But Dreser had other ideas. Salicylic acid enfeebled the heart, he announced (some doctors erroneously believed this because the high doses given to rheumatic patients sometimes made their hearts race), and acetylsalicylic acid would be just the same. He couldn't give the drug his seal of approval with that problem hanging over it. ASA was rejected.

Eichengrün was furious. A drug that was destined to become one of the most successful medicines in history was about to be consigned to the dustbin. But Dreser, in his usual infuriating way, was immovable. In any case, all his attention was taken up with another Hoffman discovery at the time, one that he genuinely believed had much greater therapeutic and commercial potential. That drug was heroin.

Like ASA, diacetylmorphine (to give heroin its correct chemical name) was not a new substance as such. It had originally been discovered in 1874 by an English chemist called C. R. Alder Wright. He had been conducting experiments with opium derivatives at St Mary's Hospital in London in 1874 and had obtained the base for the white crystalline substance by boiling up morphine in water. Curious to see what its effects might be, he tried it out on his dogs but is said to have been so disgusted by what it did to them that he threw it away. None the less he had written up the experiement and although it had since been forgotten, Dreser had come across it in one of his periodic trawls through the old scientific literature. Morphine had long been used as a painkiller and more recently in the treatment of respiratory diseases like tuberculosis, which were widespread at the time. Another opium derivative, codeine, was also in common use because of its suppressing effect on coughing fits. But both drugs had the distinct drawback of being highly addictive. Anyone who could come up with a non-addictive derivative would hit the jackpot. Knowing that acetylation (the same process used in producing acetylsalicylic acid) made certain products less toxic, Dreser figured that diacetylmorphine might just prove to be that non-addictive substance. Unusually, because it wasn't his normal role to task scientists in the pharmaceutical section, he had asked Felix Hoffman to try to replicate Wright's process. Two weeks after he had formulated ASA, Hoffman successfully synthesized diacetylmorphine (in the process earning the curious distinction of 'discovering' in the same fortnight one of the most useful substances known to medicine and one of the most deadly).

The more he had experimented with it, the more Dreser became convinced that diacetylmorphine had huge commercial potential. He began testing it on laboratory frogs and rabbits, and then on himself and some volunteers in the adjacent Bayer dye factory. It proved successful in every case – indeed, the workers had found that it made them feel so 'heroic' that the obvious brand name had suggested itself. Heroin was put out to further clinical trials and by 1998 Dreser was telling the Congress of German Naturalists and Physicians that it was ten times more effective as a cough remedy than codeine but with only one tenth of its toxic effects. A completely non-habit-forming and safe family drug, it would solve the problem of morphine addiction whilst at the same time be efficacious in a range of other conditions. The company had plans to promote it to physicians as a remedy for baby colic, colds, influenza, joint pain and other ailments – even as a general pick-me-up (much as Coca Cola was promoted in the first part of the twentieth century).

Getting this wonderful new drug tested and ready for manufacture was a lengthy and time-consuming process, none of which left much of anyone's energies for Hoffman's other formulation of ASA; even Carl Duisberg was caught up in the excitement of it all. So Eichengrün took matters into his own hands.

After trying ASA on himself, and finding it had no apparent effect on his heart, Eichengrün sent off small quantities to Felix Goldmann, Bayer's representative in Berlin. Goldmann had good contacts among the doctors there and Eichengrün asked him to arrange some discreet trials. Goldmann did as he was asked and spread it out among hospital physicians, general practitioners and even a dentist or two. Within weeks the doctors were returning glowing assessments. Not only was ASA free of the unpleasant side effects associated with salicylic acid, it also appeared to have another remarkable property – it was a general-purpose analgesic. One of the dentists had given it to a patient with toothache. The patient was hardly out of the chair before he had said, 'My toothache's gone!'

Feeling vindicated, Eichengrün circulated a report among the laboratory staff. Dreser – no doubt furious that his colleague had gone behind his back – scribbled, 'This is the usual Berlin boasting. The product has no value' in the margin, but when Carl Duisberg saw it he was intrigued and immediately ordered another full set of trials. Yet again the responses were glowing and this time, after conducting further rigorous evaluations (even trying it out on goldfish), Dreser accepted the inevitable. The drug would go into production.

On 23 January 1899, a memo circulated through Bayer's senior management addressing the thorny issue of what the new product should be called. Round robins of this sort were common and allowed everyone to have their say on a range of proposals. Because salicylic acid (as Karl Lowig had found out many years earlier) could be obtained from the meadowsweet plant, the document contained the suggestion that an abbreviation of the plant's Latin genus, *Spiraea*, should be put at the heart of the new brand name. The letter 'a' could be added at the front to acknowledge acetylation, and the letters 'in' could be tacked on to the end to make it easier to say – as was customary with many medicines of that time. It was noted that there was a drawback to this proposal because it might be suggestive of the word 'aspiration', which wouldn't have been an appropriate metaphor. An alternative could be the name 'euspirin'. When it came to him, Arthur Eichengrün, whose idea the final name probably was, wrote: 'I am in favour of Aspirin because "Eu" is generally used for improved taste and odour.' Carl Duisberg, Felix Hoffman and Heinrich Dreser all signed without comment.

It was the baptism certificate of a wonder drug.

Later that year, Dreser did his duty and wrote a pre-launch paper extolling the remarkable therapeutic benefits of the new treatment. Titled 'Pharmacological facts about Aspirin (acetylsalicylic acid)' it was the new drug's first step into the limelight. To be fair, the paper is rightly acknowledged as a scientific classic, a wonderful exposition of the medicine's chemical composition, test history and benefits. It probably did more than anything else in the early days to bring the new drug to the attention of doctors and pharmacists and played a hugely significant part in its subsequent success. But sadly, in what can only be seen as a fit of pique at the way he had been manoeuvred into approving the drug, Dreser completely omitted to mention the contribution of Eichengrün *and* Hoffman.

If Dreser was that disgruntled, he had the last laugh. Hoffman and Eichengrün received royalties only on medicines that were patented in Germany. Unfortunately, German patent law covered new processes, not new products and, as would become clear in the following years, aspirin was not deemed sufficiently new to qualify. Dreser, on the other hand, had negotiated a special deal that ensured he was paid royalties on all medicines tested in his lab. While his colleagues got nothing, he was about to become very rich.

When it finally came in July 1899, aspirin's launch was an epochal event – the culmination of an extraordinary process of investigation, chance and enterprise stretching back for centuries. But the story of this remarkable drug had only just begun.

BOOK TWO

5

PATENTS, PATIENTS, AND SELL, SELL, SELL!

ASPIRIN ENTERED THE world with more of a whisper than a bang.

In the late summer of 1899, a few hundred doctors, hospitals and pharmacists across Germany and Europe received small packages through the post. The accompanying letters explained that the parcels contained a new preparation from Farbenfabriken vormals Friedrich Bayer & Company of Elberfeld. It was an effective remedy for acute rheumatic fever and inflammation that lacked most of the unpleasant gastric side effects associated with the other standard treatments of the day – salicylic acid and sodium salicylate. It also showed some promise as an analgesic. Perhaps the recipients would like to try it out on their patients and publish any findings about its usefulness?

And that was it, apart from Heinrich Dreser's excellent paper, a scientific conference or two and a few simple announcements in the medical journals that listed the new product alongside Bayer's other drugs: heroin, Sulfonal and Phenacetin. As major product launches go, aspirin's very first appearance in the marketplace was remarkably low-key.

Yet within fifteen years aspirin was one of the most widely used drugs in the world, a pharmaceutical superstar that was outshining its rivals and redefining the uneasy relationship between medicine and commerce. Its rise to fame would be a complex interwoven process of patent battles and patent medicines, of medical ethics and advertising, counterfeiters, industrial rivalry, secret agents and competing national interests. Against this background its makers sought to exploit the drug's full commercial potential and make as much money from it as they could. They were almost too successful. What was born as a useful little medicine had, by the time it reached adolescence, become an item of strategic importance, squabbled over by ambitious men and powerful corporations. For a drug with supposedly few side effects, its historical impact would be huge.

Perhaps its eventual fate could have been guessed at from its initial reception. Aspirin may have slipped quietly on to the market, but it was very quickly noticed. The first doctor to get hold of the drug was a physician called Karl Witthauer who took part in the pre-launch clinical trials. He later confessed to have been sceptical about aspirin because new medicines appeared on the market every day and very few of them lived up to the promises of their manufacturers. But this time he was deeply impressed. He gave the drug to fifty patients at the Deaconess Hospital in Halle and reported that it had 'never failed in its effect on pain, inflammation and fever'. Another pre-launch trialist, Julius Wohlgemut, was equally keen and noted that the new compound had much stronger analgesic effects than ordinary salicylic acid. Witthaver and Wohlgemut's papers set the tone for almost all the subsequent reports. As doctors began putting Bayer's little packages to work, the word spread throughout the medical community that aspirin was a drug to be taken very seriously. More doctors were persuaded to try it and more of them gave it their blessing. Within three years around 160 scientific articles had appeared in its favour – an astounding response to a new drug, even by today's standards. Sometimes, the enthusiasm of these new disciples even outdid that of aspirin's inventors. It was more than just an anti-rheumatic treatment, they said. It was a powerful remedy for a range of other conditions too – headache, toothache, neuralgia, migraine, the common cold, influenza, 'alcoholic indisposition', tonsillitis, arthritis, perhaps even hay fever and diabetes. And, of course, the more flamboyant these claims were, the more other physicians began to prescribe aspirin and the higher its sales began to soar. In Germany, the rest of Europe and indeed the rest of the world, Bayer soon had a massive commercial success on its hands. The big question now was how best to control, protect and exploit it.

One of the company's first acts on deciding the drug should be put into production had been to try to file patents on what it saw as its intellectual property. But there was a problem. To Bayer's delight, the German patent office initially agreed to accept a patent on ASA but within a few weeks it did a volte-face, reminding the company that German law covered only new processes and not new products. ASA had actually been discovered some years before, the examiners said, if not by Charles Gerhardt, then certainly by Kolbe, Kraut et al. Regrettably, Bayer's 'discovery' could not be considered the result of a new process. Patent denied. This cannot have come as any great surprise to Carl Duisberg and the Bayer board who knew the ins and outs of the German patent system as well as anyone, but for a time they had higher hopes for elsewhere in the world. However, it soon became apparent that most other countries took the same view – ASA was not the result of a new process and new products alone didn't qualify. It seemed that only two nations would grant Bayer the protection it sought. Fortunately, they were also the two largest potential markets: Britain and the United States of America.

The British patent was filed on 22 December 1898, even before the new drug was named. Accepted under the number 27,088, it granted protection to one Henry Edward Newton for 'the manufacture of production of "acetylsalicylic acid" communicated to him from abroad by the Farbenfabriken vormals Friedrich Bayer & Co. of Elberfeld, Germany'. The said Mr Newton was just a nominee, of course, a Bayer licensee whose name was placed on the document because of his British nationality. The patent specification contained the justification that, 'My foreign correspondents [Bayer & Co.] have now found that on heating salicylic acid with acetic anhydride, a body is obtained, the properties of which are entirely different from those of the body described by Kraut . . .'

The same justification was inserted into the US patent filed on 27 February 1900 which began: 'Be it known that I, Felix Hoffman, doctor of philosophy, chemist (assignor to the Farbenfabriken of Elberfeld Company of New York) have invented a new and useful Improvement in the Manufacture or Production of Acetyl Salicylic Acid . . .'

As it happened, both the justification that Bayer's process was different to Kraut's, and the fact that Hoffman's name was put on the US patent, would later cause no end of problems. But for the moment, all appeared well. In two of the biggest markets in the world, Bayer had managed to get a monopoly (albeit a time limited one) over the production and sale of what was clearly a very popular new drug. And of course, it had been able to obtain trademarks on the aspirin name everywhere, because it was genuinely a new word. Trademarks, as Carl Duisberg remembered about Phenacetin, could be even more powerful than patents if properly exploited. If a manufacturer could irredeemably associate the brand name of a product with an action or a quality in the mind of a consumer, that consumer would return time after time to the product, no matter how effective or attractive an identical (but differently branded) offering from a rival might be. The only question was how best to make that link.

And thus Bayer faced its biggest conundrum, the ultimate resolution of which would have such a significant effect on its fortunes and on the fate of aspirin. It was a problem perhaps best summed up by three words: *Lily the Pink*.

Dr Ryan's Incomparable Worm-Destroying Sugar Plums, Bardwell's Aromatic Lozenges of Steel, Darbey's Carminative, Turlington's Balsam of Life – the kinds of nostrums and elixirs that the Reverend Edward Stone might have found on a market stall in mid-eighteenth-century Oxfordshire still flourished 150 years later. For all the many advances in scientific medicine that had been made in the intervening years, there were still enough gullible people to rise to the bait of the miracle-sellers and travelling hucksters. In fact, technology, in the shape of newspapers and the railways, had given the nostrum-monger access to more people than ever before. Though increasingly seen as down-market and old-fashioned by the 'respectable' middle classes, the quack remedy business had kept a firm grip on the hearts, minds and wallets of the poorer and more credulous. Along the way their products had become known as patent medicines – after a term coined in England around Edward Stone's time when patents were often used to prevent people finding out just how useless the ingredients of quack remedies were. And, of course, the claims had, if anything, become more outlandish.

Take this typical (*see right*) advertisement in the *New York Times* of 3 May 1887.

Lydia Pinkham of Lynn, Massachusetts, started peddling her Vegetable Compound around small-town America in 1873 and was so successful that she soon became one of the country's first women millionaires. Sadly, her miraculous elixir didn't seem to do much for her own health. She suffered a stroke and died prematurely in 1883. This failed to deter her business partners, however. For many years after her death the company continued to publish advertisements under her name as though she were still alive, and every one of the millions of bottles sold carried a picture of Lydia in beaming good health. Even more remarkably, she still somehow managed to pen replies to women who wrote in to the company asking for medical advice. It seemed that people couldn't swallow enough of her secret formula or the guff that went with it, any more than they could resist other exciting remedies of the day like Dr Wright's Indian Vegetable Pills or Steer's Genuine Opedeldor. Perhaps in the case of Lydia's remedy this popularity owed much to the fact that its most active ingredient was alcohol – something that was kept quiet at the time. Many years later she would be ironically celebrated in song as 'Lily the Pink, the saviour of the human race', but the sad truth was that her customers were more likely to get a hangover than anything else.

They weren't the only ones getting a headache from patent medicines. By the end of the nineteenth century these had become a sore trial to the medical and pharmacy communities too. Pharmacists were now being scientifically trained and saw themselves as experts with a professional creed. Although they had once been happy to make money selling dodgy remedies to gullible laymen, this had gradually become less and less acceptable as time went by. Physicians had also learnt how to distinguish between what was good and what was spurious. In the old days prescription had been a simple matter – the patent medicines were often no better or worse than the traditional treatments found in the professional pharmacopoeia, and so allowing patients to take them seemed harmless enough. But as legitimate scientific treatments improved and the failings and hazards of patent medicines became more obvious, so standards had become more rigorous. A heated debate began about how these professions should respond to quack remedies, particularly in the United States where advertising, and hence public demand, were especially fierce. The American Medical Association and the American Pharmaceutical Association eventually made it clear that substances protected by patent, secrecy or trademarks were not to be officially sanctioned by prescription. The public needed to know whether or not a medicine could be trusted. If it could, then everyone should be able to have it, regardless of its ownership. If it couldn't be trusted then it couldn't be recommended.

Guidelines like these (similar rules were introduced across Europe at around the same time) did not have much impact on the new chemical industries at first, because they did not deal directly with the public. The substances they produced were only sold as raw ingredients for professional pharmacists to make into finished medicines, and because those pharmacists had learnt how to analyse the quality of the ingredients scientifically, they were able to guarantee that the final compounded product was as safe and as effective as possible. As a result, doctors had learnt to trust the pharmacists to prepare remedies in accordance with their prescriptions, secure in the knowledge that nothing harmful to the patient would get past them. The net result were so-called ethical drugs, much like today's prescription drug: products available only from a pharmacist on a doctor's order and distinct from the patent medicines which could be bought over the counter in any small grocery store.

But when the German chemical companies began making their own finished synthetic medicines this cosy relationship came under strain. They continued to regard themselves as ethical producers because they still restricted their sales to the medical professions. They were also in business to make money, however, and were keen to use the same marketing tools that had proved so successful in their previous incarnation as dye producers: namely, trademarks, patents and advertising. Unfortunately, to the pharmacy profession this whiff of hard commerce smelt suspiciously reminiscent of the tactics used by the quack remedy industry. And when pharmacists began to complain, the physicians listened.

Their heightened sensitivity had been the main reason behind Bayer's low-key approach to the launch of aspirin. The company needed the good will of the medical and pharmaceutical professions for the drug to succeed as they would be prescribing and dispensing it. Trumpeting the attributes of the new product too loudly – or too commercially – risked jeopardizing that support.

However, the company soon realized that it had a potential gold mine on its hands and wanted to exploit it for all it was worth. That meant pushing the boundaries of what was acceptable and using its whole armoury of commercial weapons. It was clear that it would be particularly important to do this in the United States – one of only two countries where Bayer had secured a patented monopoly on acetylsalicylic acid and, as its largest potential market, the country where it had most to gain. One day the US patent would run out and rival manufacturers would then be free to pounce. Before that happened the aspirin brand name had to be anchored firmly in the American consciousness so that when customers wanted acetylsalicylic acid, they thought automatically of the Bayer product. Unfortunately, America was the place where brash commercialism in the drug business was now most frowned upon by the medical establishment. What was Bayer to do?

It was a dilemma that gave Carl Duisberg many sleepless nights. But he knew that whatever the solution might be, it could not be found until other problems facing aspirin in America had been resolved. Bayer might have had a potentially huge success on its hands, but unless the company was able to secure and protect its wider competitive position in the United States, that success could be snuffed out before it could be properly exploited.

The firm's American business was conducted through a subsidiary, the Farbenfabriken of Elberfeld Company, set up in the late 1860s to handle US sales of its dyestuffs and other chemicals. These had always made a healthy contribution to Bayer profits, but the move into pharmaceuticals had been much more problematic. Even though the company had won a US patent on its first big drug, Phenacetin, high import duties had made the product an attractive target for smugglers who bought it cheaply in Europe and slipped it through Canada and Mexico on to the American black market. Duisberg had bombarded the company's US-based executives with instructions about how to stop this trade, from hiring more salespeople to pursuing the illegal importers through the courts, but the loss of revenue to Bayer had still been considerable. Indeed, Duisberg knew that the situation was likely to worsen in 1906 when the Phenacetin patent expired and legitimate American manufacturers would be free to sell it cheaply as well.

He was determined to prevent the same fate for aspirin and so in 1903 he set sail for America to explore a possible solution. If Bayer drugs could be made in the United States rather than Germany they would be tariff-free. The price to consumers would thus be brought down and bootleggers and any future legitimate rivals would be deprived of their competitive advantage. Duisberg didn't much like the loss of central control this would necessitate – as an arch centralist he found the idea of a semi-autonomous US operation rather worrying (a fear later justified by events). But if that was what it would take to protect his new wonderdrug, then so be it.

Fortunately, Bayer already had some manufacturing experience in America. Its subsidiary owned a modest stake in the small Hudson River Aniline and Color Works at Rensselaer in upstate New York. The factory had good communications links and a pool of immigrant German labour in nearby Albany. Duisberg realized that if Bayer acquired the rest of the company and invested in extra plant and facilities for pharmaceutical production, Rensselaer could be the answer he was looking for – an American home for aspirin. And so the money was found and a new factory was built – one of the largest and most up to date in the country.

With that resolved, Duisberg and his colleagues were free to turn their attention back to their other thorny dilemma – how to exploit the huge potential of the US patent without alienating the anti-commercial zealots of the American medical profession. But just as they began to work out a bold new marketing strategy, a further unexpected complication loomed – this time from Britain.

Shortly before eleven on the morning of 2 May 1905, George Moulton KC stood up in the majestic oaken surroundings of London's High Court and prepared to fire the opening salvoes in one of the most significant intellectual property battles in medical-legal history. Alongside and behind him sat a bewigged supporting cast, like him, all acting for the plaintiff, Farbenfabriken vormals Friedrich Bayer & Company. On the other side of the room sat a similarly impressive legal team for the defendants: Chemische Fabrik Von Heyden. Huge bundles of documents, neatly tied with red ribbon, lay on the tables in front of everyone, and at the back, waiting nervously to be called, sat some of Europe's leading experts in the arcane new field of pharmaceutical chemistry. All of this was for the benefit of His Lordship Mr Justice Joyce, one of the most experienced judges in the Chancery Division. Perched in lonely authority on a high bench at the head of the court, he peered down at the plaintiff's chief barrister and signalled that he should begin.

'My Lord,' said Moulton, 'this is an action for infringement of patent number 27,088 . . .'

That two of Germany's leading chemical companies should have been slugging it out in a British court might seem surprising at first, but the stakes were enormous. A UK patent was a thing of huge value at the time. It didn't just protect the patentee from competition in Britain, it also applied to many of the dominions and territories in its sizeable empire, from India in the East to Canada in the West. In reality it was impossible to police this writ in the farthest-flung parts of the world, but foreign companies that wished to sell into those countries were still better off having one because of the influence the mother country exercised over trade. A British patent was like a certificate of authenticity that demonstrated the value of the goods. And of course the market for those goods in Britain alone was potentially enormous.

The issue in this particular case was Bayer's patent for acetylsalicylic acid, filed in 1898, and the company's claim for damages over what it alleged was a clear case of infringement by Chemische Fabrik Von Heyden. Behind the cold legal language of its writ, Bayer's spluttering indignation was there for all to see. That these arrivistes, these upstarts, had had the temerity to start selling acetylsalicylic acid in Britain, when they knew full well that ASA was a Bayer discovery, patented as such and protected by the full force of British law, was a gross outrage that had to be punished immediately.

Of course, as many of the chemistry experts in the court also knew full well, this was actually an old, old rivalry. In fact, if anyone could be accused of being an arriviste it was Bayer. Chemische Fabrik Von

Heyden had been in this game for many years – since 1859 to be exact. That was when the Marburg University professor Hermann Kolbe had managed to synthesize salicylic acid from sodium phenolate and carbon dioxide. His student Friedrich Von Heyden had taken this process up commercially and had founded Chemische Fabrik Von Heyden to produce it. Since then Heyden had become one of Europe's leading producers of salicylic products and in 1901 had brought out its own unbranded version of acetylsalicylic acid – the substance Bayer now sold as aspirin. Bayer had been unable to do anything about it in Germany because ASA wasn't protected by a patent there, but when Heyden began importing the drug to Britain, Bayer's lawyers had pounced. They bought 2 lb of Heyden ASA, tested it, declared it to be an identical formulation to their own, and filed suit.

Chemische Fabrik Von Heyden's response had been cool and unexpectedly frank. Yes, it admitted, it had been importing ASA to Britain and yes, that substance was virtually identical to the Bayer product. But Bayer had obtained a UK patent under false pretences. It claimed to have made a new discovery, whereas in actual fact acetylsalicylic acid had been discovered first by Charles Gerhardt some fifty years before. The process had then been refined several times over by Professors Kolbe, Johann Kraut and others too numerous to mention. In other words, Bayer's patent was a fake that should never have been granted. And if it should never have been granted, Chemische Fabrik Von Heyden couldn't possibly be guilty of infringing it.

It was a neat defensive argument, and one that spelt potential disaster for Bayer. One slip and it would be open season on the hottest pharmaceutical product in years. But the company had no choice but to fight it aggressively. In any case, Bayer was confident that its patent would be upheld. Others might have come up with versions of ASA, but until Felix Hoffman, no one else had been able to produce the pure product. That was clearly the case. Wasn't it?

Hence the battery of expensive lawyers, the huge piles of documents, the ranks of eminent witnesses and the learned attention of His Lordship Mr Justice Joyce, who had taken the unusual precaution of bringing a basic chemistry textbook into court with him.

He needed it. Over the next eight working days he was bombarded with facts and figures, chemical formulae, legal precedents, lengthy articles from German scientific journals and hours of expert but often conflicting testimony from some of Europe's leading chemistry professors, pharmaceutical scientists and doctors. His Lordship sat impassively throughout, stopping proceedings only to ask an odd question here, clear up an odd point there, or rifle through the pages of his chemistry primer in search of a definition.

As is often the way in such cases, the arguments eventually focused down on to one point of law. Bayer's patent contained the justification that the properties of its product were 'entirely different from those of the body described by Kraut . . .' So were they different or not? From their places in the public gallery, rival German businessmen and their British representatives tried to work out how Joyce was responding to the arguments. Did he understand the science? Was he convinced by the testimony of Sir James Dewar and Dr Adolf Liebmann for the plaintiffs, or were the statements of defence witnesses Frankland, Armstrong and Rosenheim more compelling? Which way would the judge go? Joyce gave them no indication. On 11 May, he took closing arguments from counsel and then announced an adjournment while he went away and pondered his decision. As everyone filed out of court, their heads spinning with descriptions of laboratory techniques, the rival lawyers murmured quietly reassuring words in the ears of their clients and went back to their chambers to tot up their huge fees. Whatever the outcome, one thing was already clear: aspirin was going to be a very profitable product for the legal profession.

On 8 July, they were all back in court and this time Joyce had something to tell them. He began with a succinct summation of the case, betraying an understanding of pharmaceutical chemistry that would have done credit to many of the professional scientists in the room. Then he got into his stride and the Bayer representatives must have wished the courtroom floor would open and swallow them up.

The Bayer patent, he said, was a remarkable document, the like of which none of the experienced counsel in the case had ever seen before. It was 'erroneous and misleading . . . by accident, error or design so framed as to obscure the subject as much as possible'. He continued remorselessly,

> It would be a strange and marvellous thing, and to my mind much to be regretted, if after all that had been done and published with regard to acetyl salicylic acid before the date of this patent, an ingenious person, by merely

> putting forward a different, if you like a better mode of purification from that stated, and truly stated by Kraut to be feasible, could successfully claim as his invention and obtain a valid patent for the production of acetylsalicylic acid as a new body or compound. In my opinion, it was not a new body or compound and I hold the patent in question in this case to be invalid.

With a final 'this specification contains no element of invention or discovery beyond what was common knowledge . . .' Joyce awarded this 'very peculiar case' to the defendants. As Heyden's lawyers entered an immediate and successful plea for full costs, the Bayer team sat in stunned silence. One can imagine what was going through their minds. Who was going to tell Carl Duisberg?

The shockwaves from Mr Justice Joyce's decision were considerable. In a wider sense his comments about the obscurity of the language used in the Bayer patent were symptomatic of a growing unease at that time about the stranglehold that German chemical companies had on the secrets of their trade. On both sides of the Atlantic domestic businessmen had begun to complain that 'foreigners' were abusing the patents process.

To be fair to companies like Bayer, this approach was born out of their experience with their own German system and its loopholes that allowed a company to produce someone else's product if they could come up with a slightly different way of making it. Obscuring the methodology behind an invention to make it more difficult to copy was something that all German companies were used to doing, and it was only natural that they followed the same procedures in Britain and the United States. But it didn't make them many friends. Bayer's loss of its British aspirin patent was seen by contemporary commentators as a case of 'he who lives by the sword . . .' It served the company right, in other words.

The decision led indirectly to a reform of British patent laws. David Lloyd George, who would later be the country's wartime Prime Minister, had just become the Minister of Trade and would soon begin drawing up the necessary legislation. On its introduction, two years later, he used words that were clearly influenced by Mr Justice Joyce.

> Big foreign syndicates have one very effective way of destroying British industry. They first of all apply for patents on a very considerable scale. They suggest every possible combination, for instance, in chemicals, which human ingenuity can possibly think of. These combinations the syndicates have not tried themselves. They are not in operation, say, in Germany or elsewhere, but the syndicates put them in their patents in obscure and vague terms so as to cover any possible invention that may be discovered afterwards in this country.

Such sentiments struck a chord with British manufacturers who had been saying much the same thing for some years. But perhaps more importantly, they also reflected a wider malaise in international relations. The more people in high places hinted at foreign conspiracies and plots like this, the more polarized popular opinion became. In Germany, it was coalescing around those who believed that their nation was being denied a place at the industrial and political top table, whilst the Anglo-French camp feared that German muscle-flexing was beginning to threaten their own security. These were the days of Erskine Childers and *The Riddle of the Sands* (1903), of Dreadnought launches and invasion fears. Europe's yellow press was stoking up suspicion and the politicians were going along with this. In such a febrile atmosphere, German pre-eminence in the chemical sciences began to assume a significance beyond the merely commercial. It was now seen as a matter of strategic importance as well. The question of who held which patents to which technologies had hitherto been something that only lawyers and businessmen worried about. But in the darker days to come, even a comparatively minor dispute over a German chemical company's new drug would be remembered and weighed in the balance. Joyce's decision would have much wider implications than anyone could have foreseen at the time.

Of course, in the short term, the court case had most effect on Bayer itself. In Britain it forced the company back on to the strength of its aspirin trademark. Now other companies could import ASA (the British chemical industry wasn't yet up to producing it), the brand name was the only thing left to distinguish the Bayer product from that of its rivals. As a result it would have to push that brand name with all its might, even though it increased the risk of direct conflict with the anti-commercial lobby in the British medical profession, many members of which felt as passionately about the issue as their counterparts in America.

In Germany, it gave a considerable boost to Bayer's rivals in the synthetic drug trade. Chemische Fabrik Von Heyden wasn't the only competitor making ASA and for some time there had been complaints among German pharmacists that Bayer's aspirin was considerably more expensive than the ASA available from other producers. A number of experimenters had compared the purity of different makes of ASA with aspirin and had found the leading unbranded products to be almost identical to Bayer's. Now that the huge British market was open to competition, the stakes were large enough to encourage Bayer's rivals to try to take bigger bites from its domestic market too. It's the same product, they argued. Even the British have recognized that. So why pay more than you have to? Once again, the only way to meet this challenge was by more assertive marketing of the aspirin brand name.

In the United States the decision had even more dire implications. American courts frequently followed precedents set in Britain and, as it happened, shortly before the British patent case had come to trial, Bayer had filed a similar infringement suit in Chicago. When Carl Duisberg and the Bayer board sat down to consider the possible repercussions of their lost British patent, this case was uppermost in their minds.

The problem was that Bayer was the victim of its own success in the United States. To get round the hostility of the American Medical Association (AMA) towards anything that directly created patient demand for particular medicines, the company was following the ethical route of promoting aspirin only to doctors. But it had been very aggressive in doing so. Its salespeople had been banging on surgery doors across America, flooding physicians with free samples and copies of the many learned articles in praise of aspirin. They had even advertised, albeit discreetly, in the *Journal of the American Medical Association*, the physicians' own house journal. And the strategy had worked. By 1906 it already accounted for 25 per cent of the company's total sales in the United States and soon Bayer was claiming:

> Aspirin has in the decade since its introduction become so popular that it is unsurpassed by any other drug. Surely, it is not an exaggeration to say that it is today the most used and most beloved medicine that we manufacture.

But just as had happened previously with Phenacetin, this huge success had encouraged smugglers to start shipping in cheaper ASA from abroad – a trade that hadn't yet been dented by the cost savings from the new Rensselaer plant. Even worse, back-street counterfeiters had started to produce impure ASA, often labelled as 'aspirin', that unwitting doctors (not up to making their own chemical analyses) had been buying in good faith as the genuine product. More worryingly still, some of this fake aspirin was being sold direct to the public.

So Bayer had decided to make an example of one of the worst patent infringers, a Chicago pharmaceutical wholesaler called Edward A. Kuehmsted. The plan was to drive him out of the bootleg business, and if possible bankrupt him, thus serving warning on other infringers and counterfeiters that any abuses of Bayer's monopoly would be met with fierce reprisals. In the process, Bayer's US aspirin patent would be backed by the certainty of a legal endorsement, making it that much easier to sue new offenders.

Of course, just as Chemische Von Heyden had done in Britain, Kuehmsted's lawyers claimed the original patent was invalid and that therefore he could not have transgressed. But Bayer was confident that its experts could take on any arguments put forward by a Chicago bootlegger – or at least it was until the dreadful news from arrived from London.

The risk of losing the case because of the British precedent meant it was now too dangerous to proceed. But the potential consequences of abandoning the suit were equally unthinkable. There was only one way out – to stall. The longer Bayer's lawyers could spin the case out, the longer the patent would remain valid and the greater the breathing space Bayer would have to exploit the drug. So that was what its lawyers managed to do for the next five years, fighting off every attempt by Kuehmsted's legal team to get a result. In the event, and much to everyone's surprise, when the case was finally resolved in 1909, the presiding judge found in Bayer's favour. But in the intervening years, Bayer hadn't been able to take this outcome for granted, and even if it had, there was always the time limit on the patent, due to expire on 27 February 1917 – a date that had begun to loom ever larger in the company's consciousness.

For Carl Duisberg and his colleagues, it had all brought into focus the question that had dogged their infant prodigy right from the start:

how to get the aspirin brand name into the hearts and minds of as many Americans as possible, before identical (and cheaper) products became freely available elsewhere. The answer was clearly the same as it was in Europe. Take on the doctors and pharmacists and sell, sell, sell! But Bayer could hardly have picked a worse time to start a new promotional push. The prickliness of the American medical establishment was about to reach new levels of intensity. And once again patent medicines were to blame.

If the American newspaper industry had been partly responsible for the commercial success of patent medicines in the nineteenth century (a charge it could hardly deny given that it had published so many advertisements for bogus remedies), it began to redeem itself at the start of the twentieth when some of its more progressive journals began to campaign against their shortcomings. The most significant of these attacks appeared in *Collier's Magazine* in 1905. Its editor, Norman Hapgood, had grown disgusted by the blatant lies he had found in advertisements submitted to him by makers of such products as Buffalo Lithia Water (a substance endorsed, apparently, by the physician-in-ordinary to the Pope and miraculously able to relieve everything from gout to dyspepsia). Hapgood decided to start a campaign against proprietaries and looked around for a suitably aggressive reporter to take it on. He found Samuel Hopkins Adams, one of a new generation of tabloid journalists who had made his name covering sensational murder cases in New York. It was to prove an inspired appointment.

In April 1905, the two sat down together and plotted strategy. Hapgood would publish a series of 'softening-up' editorials on the general menace of the nostrum trade, while Adams would go away and start digging up the real dirt. Over the next few months, as *Collier's Magazine* began its campaign, Adams travelled the country buying up patent medicines and getting them scientifically analysed, talking to government chemists and medical experts, tracking down people who had supposedly given personal 'testimonials' quoted in advertisements and generally making a nuisance of himself. None of this went unnoticed by the patent remedy business, of course, and Adams had to shake off private detectives and at least one attempt at blackmail. But he persisted, and on 7 October 1905, his series began. It was called 'The Great American Fraud' and his opening words set the scene for what was to come.

> Gullible America will spend this year some seventy-five millions of dollars in the purchase of patent medicines. In consideration of this sum it will swallow huge quantities of alcohol, an appalling amount of opiates and narcotics, a wide assortment of varied drugs ranging from powerful and dangerous heart depressants to insidious liver stimulants; and far in excess of all other ingredients, undiluted fraud. For fraud, exploited by the skilfulest [sic] of advertising bunco men, is the basis of this trade. Should the newspapers, the magazines and the medical journals refuse their pages to this class of advertisements, the patent medicine business in five years would be as scandalously historic as the South Sea Bubble, and the nation would be the richer not only in lives and money, but in drunkards and drug-fiends saved.

Over the next ten weeks Adams weighed into the patent medicine business, its fraudulent claims, its bogus medicaments and grotesque profits. It was a *tour de force* of investigative journalism that left no stone unturned and, not surprisingly, it caused a public sensation. But most important of all was the boost the series gave to the quack trade's greatest enemy – Harvey Washington Wiley.

Born in 1844 on a small farm in southern Indiana, Wiley was a quite remarkable man. He had qualified as a doctor after a brief spell in the army during the Civil War, but decided to devote himself to the nutritional side of health rather than practise as a physician. After a brief academic career he went into public service as Indiana's chief chemist. It was there that he discovered a passion for food analysis and a hatred for anything and anyone who adulterated it. He took both with him to Washington, DC, in 1883 when he became head of the Department of Agriculture's Bureau of Chemistry.

A large, bull-like figure with unruly hair and a rough, clubbable charm, Wiley was a crusader and an innovator. One of the first people in Washington to drive a car and reputedly the first to crash one, he set about transforming the federal government's procedures for investigating the purity of the nation's food supply. His small bureau tested thousands of different foodstuffs, revealing for the first time how creative chemistry was being used to doctor the taste and appearance of almost everything America ate. The more scandals he exposed, the more pressure he put on Congress to do something about it. The press picked up on his reports and resulting publicity created a touchstone issue for the new progressive government of President Theodore Roosevelt.

Before long Wiley had turned his attention to the patent medicine business and become an outspoken and indefatigable champion for purity in medicines too. His attempts to address the problems facing both food and drugs kept falling foul of powerful Congressional lobbying by the industries concerned, but Wiley was ferocious in keeping the issue on the agenda. He became a member of the AMA's Council on Pharmacy and Chemistry and used these connections to press for food and drug legislation. Adams had sought his help when writing his articles and their publication was exactly the sort of ammunition Wiley needed. To add grist to his mill, a few days after the last of the *Collier's* pieces appeared Upton Sinclair published his novel *The Jungle*, which described in appalling detail the squalor of the stockyards in Chicago. The combined public outcry over adulterated meat and poisonous quack remedies was gleefully whipped up further by Wiley in the political salons around Washington and eventually came to President Roosevelt's attention. He insisted that Congress finally do something about it. In June 1906 the Food and Drug Act was signed into law.

The drug portion of the law didn't cover advertising (a gaping hole that would later cause still more headaches) but it did cover labelling of drugs. For the first time the ingredients of a medicine had to be accurately described on the container in which it was sold. Soothing syrups and rheumatic remedies now had to reveal whether they contained cocaine, opium or any other potentially harmful substance.

The following year the law received its first test when Wiley instituted proceedings against a proprietary medicine brand called Cufordehake Brane Fude. This was a popular over-the-counter analgesic that contained two of the earlier discoveries of the German synthetic medicine industry – acetanilide and antipyrine. Unlike aspirin, which Bayer was selling only through doctor's prescription, acetanilide and antipyrine were years out of patent and were therefore free to be used by patent medicine makers in all sorts of unstable compounds. Acetanilide in particular was reckoned by some to have serious side effects on the liver and kidneys. The problem for the owners of Cufordehake Brane Fude was that these dangerous ingredients were not mentioned on its label. Armed with his new law, Wiley secured a conviction. The penalty was small – a fine of 700 dollars – but it sent out a loud message. The unprincipled promotion of medicines, from whatever source, had to stop.

The influence of all this activity – press exposés, legislation, Presidential interest and court cases – on relations between the American medical profession and the mainstream synthetic drug producers was considerable. For a start, it highlighted a problem that the AMA and others had been warning about for some time. Ever since the German pharmaceutical companies had signalled their intention to stay ethical by marketing and selling only to physicians and pharmacists, less scrupulous patent medicine producers had been copying their approach by selling pseudo-ethical medicines in the same way. Many doctors were unable to discriminate between these products, and several quack remedies masquerading as legitimate ethical drugs had found their way via medical journal advertisements on to prescription pads. Now, with everyone made extra sensitive by all the public attention on the dangers of patent remedies, the governing bodies of the American medicine and pharmacy professions decided enough was enough. One consequence of the Food and Drug Act was to restrict entrance to the official US Pharmacopoeia to non-trademarked drugs – a way to keep branded patent medicines off the prescription list. Henceforth, generic chemical names must also be used wherever possible. Moreover, advertising's distasteful association with quack remedies meant that any advertisement containing more than a company name and the name of the product was considered excessive. Even the synthetic medicine producers would have to fall into line.

For Bayer, reeling from the loss of its British patent and concerned about the potential loss of its American one, the fracas started by 'The Great American Fraud' gave it a whole new set of headaches. The company had nothing to do with patent remedies and bitterly resented being put in the same boat as the quacks. But how could it market a branded ethical drug when there was such hostility to trademarks around? It had to do something because the aspirin brand was vital to its future success. Indeed, it could be argued that it was vital to public safety as well. The drug had proved so popular that the counterfeit trade was mushrooming (illegal copies accounted for almost half the total aspirin sold in 1909). The problem had been compounded by Bayer's absolute determination to retain aspirin's status as an ethical drug, which meant that it had only been sold in powder form to wholesalers for processing into unmarked tablets. Consequently, although pharmacists and doctors might realize that aspirin was a Bayer product, the public didn't know and were unable to tell if they were taking the real thing or being fobbed off with a fake. Bayer had to let it be known that only genuine Bayer Aspirin would do. Its solution was to start making the tablets itself, each pill stamped with a new company logo, the famous Bayer cross – two perpendicular Bayer names that crossed at the central letter 'y'. But even this opened the company up to accusations of crass commercialism from the AMA.

Then there was the further problem that only drugs called by their generic chemical names could now be listed in the US Pharmacopoeia. Without that listing, aspirin wouldn't be prescribed, but if it could only be listed as acetylsalicylic acid in the official lists, what would stop pharmacists turning to identically named rival products when Bayer's ASA monopoly ran out? Bayer's answer was to give the drug the nonsensical generic designation 'monoacetic acid ester of salicylic acid', the theory being that few American physicians would ever remember such a complicated term or know that it was the same thing as acetylsalicylic acid. Most would continue to put 'aspirin' down on their prescription pads just as they always had. But not everyone was fooled – the influential *Druggists' Circular Chemists' Gazette* called it a 'dodge to keep just ahead of the law', and did its best to alert doctors to the plan.

And then, last but not least, there was that looming patent expiry date. Bayer had clung on to its monopoly in the aftermath of its long-drawn-out Chicago suit, and had used the result to browbeat and sue every other patent-infringer it could find, but the vultures were circling. Already the United States' own chemical industry was signalling its determination to jump in and begin producing ASA the moment the patent expired. It would be a desperate struggle to keep them at bay.

If Bayer had been an American company and Carl Duisberg an American executive, all these would have been problems enough. But, of course, they weren't. Bayer and Duisberg were German. And in August 1914, Germany went to war.

6

THE CHEMISTS' WAR

Sir,

I am pleased to see the advertisement of 'Helicon' in your issue of last week. I find that helicon and aspirin are identically the same.

Now is the time. Prescribe helicon and put money in English pockets and not aspirin and money in German pockets. This is to be a war of endurance and the length of the purse will settle it.

I am, sir, yours faithfully.

Edward Treeves MRCS Eng

Whether the jingoistic fervour gleaming through this letter to the *Lancet* in October 1914 shone quite so brightly at the end of the war is unknown, but it's unlikely. It would have been remarkable indeed if Dr Treeves and his family had remained untouched by the monstrous casualties and hardships of the next four years. None the less, it was typical of the correspondence that flooded into the British press following the outbreak of hostilities. Indeed, similar sentiments could be found in the newspapers and journals of all Europe's combatant nations that autumn. This was war and in wartime you didn't aid the enemy by buying his products – even if you could get hold of them. You looked for something closer to home.

The First World War was a watershed for aspirin. Even as the poppies began to sprout on the first shell-torn battlegrounds of the Western Front, Bayer's control over its fifteen-year-old *Wunderkind* was starting to slip. By the time the war shuddered to a close, others would be claiming the drug as their own. Aspirin, they would declare, was too important and too profitable a discovery to be meekly handed back to the losers, and in any case, they had learnt how to make it for themselves.

The process began in Britain. Trade with Germany ceased with the outbreak of hostilities and in the months following August 1914 the assets of German companies still left in the country were frozen or taken over. German citizens and immigrants who had been unable or unwilling to return home were either interned or forced to make an uncomfortable choice as to where their loyalties lay. Of course, the same thing was happening to alien nationals and assets in all of Europe's combatant states. It was more than just a matter of disrupting the other side's trade, said their governments. National interest dictated that strict control was kept over commodities of strategic importance and that the enemy was denied any use of them. As Dr Treeves had written so presciently, this was to be a war of endurance and deep pockets would be decisive.

Britain had one big advantage in this respect – superiority at sea. Although it would be seriously tested at times by Germany's new weapons – U-boats, torpedoes, surface raiders and magnetic mines (and more flamboyantly, if less effectively, by the enemy's battleships) – the Royal Navy was able to control events at sea to far greater an extent than the Allied armies struggling in the Flanders mud. And right from the start, a key part of its strategy, just as it had been against Napoleonic France 100 years earlier, was the enforcement of a blockade against enemy shipping. This was being put into place even as the first British troops were being ferried across the Channel and was soon, and controversially, being applied to Germany's trade with neutrals as well.

Unfortunately, this trade embargo was something of a double-edged sword for Britain because it cut her off from those goods that only German had hitherto been able to provide. Britain's Allies and her empire might be able to supply the extra fuel, food, textiles and munitions that she needed but there were numerous other commodities that were not so easy to find. The products of the mighty German chemical industry were no longer available, for instance, and although domestic producers gradually picked up the slack, there were some things like pharmaceuticals that were much harder to replicate. Several important drugs were going to be in short supply.

Aspirin was one of those drugs. Bayer had lost its British patent to acetylsalicylic acid back in 1905, allowing others either to import it or make it domestically. But although rival products had soon appeared on the market, most of the non-Bayer ASA sold in Britain had continued to be German in origin – from companies like Heyden and Hoechst. Furthermore, Bayer, through its British subsidiary, the Bayer Company Ltd, had retained its rights to the aspirin brand name and had been fiercely promoting and protecting it. The name had eaten its way into British minds as deeply as it had elsewhere in the world. When doctors in London or Manchester prescribed ASA, most of them wrote 'aspirin' out of force of habit, and their patients had come to assume that aspirin was a drug in its own right – just as Bayer had intended them to do. As a result, it had been hard for other ASA producers to break through this stranglehold on the market, and of course what they couldn't hope to sell, they hadn't bothered to make. When Bayer and the other German producers stopped exporting the drug, Britain's domestic pharmaceutical industry was not immediately geared up to replace them. The best-selling locally manufactured brand was called Xaxa, made by Burroughs Wellcome and there were others like Dr Treeves's Helicon, but prior to 1914, their sales had been tiny in comparison.

This shortfall was compounded by the fact that there was only so much chemical expertise and manufacturing capacity to go around. When war broke out, much of Britain's domestic chemical and pharmaceutical industry had to switch over to the manufacture of other, more deadly products. And although, as the British Pharmaceutical Society reminded its members, producing ASA in the laboratory was theoretically quite easy, making it industrially was more complicated; not least because there was a dire shortage of salicylic acid from which ASA was compounded. One of synthetic salicylic acid's key ingredients was the chemical phenol, but phenol was also vital to the manufacture of high explosives and most of the limited national supply had already been ring-fenced by the armaments industry.* Until this shortage was remedied, the more shells that British guns fired, the less British ASA could be produced.

The British government appreciated the problem, but then in the first months of the war it faced many thousands of similar ones. It was not until 5 February 1915 that it got around to taking action. The announcement was carried in the *Lancet* later that month.

---

* Phenol is otherwise known as carbolic acid and was the basis of the disinfectant that the nineteenth-century surgeon Joseph Lister had used to reduce post-operative infections.

> [The Aspirin Trade Mark]
> The Board of Trade has made an order regarding the aspirin trade marks, the effect of which is to make the word 'aspirin' public property. As is well known, this is the trade name under which the Bayer Company introduced acetylsalicylic acid and the name under which the drug is best known. It is now open to anyone to sell acetylsalicylic acid as aspirin and it is to be hoped that this will not lead to any deterioration in the standard of the drug . . . there are now several makers of acetylsalicylic acid in this country, and an examination of various samples has shown that chemically most are identical with the original aspirin.

The news was greeted with patriotic glee in the trade press, with much harking back to Mr Justice Joyce's decision to renounce the Bayer ASA patent in 1905. One journal boasted that, 'The large army of aspirin consumers will now receive a British product, and another stronghold of the enemy be destroyed.' But geeing up potential manufacturers by allowing them to call their product aspirin was one thing, getting them to make it in sufficient bulk and quality was another. Whoever the *Lancet*'s 'several makers of acetylsalicylic acid in this country' were, it seems that few of them had the skills or resources to do the job properly, although they were happy to take the profits. The sad truth was that much of the new ASA product wasn't chemically identical to aspirin at all; it was woefully impure and often unpleasant to take. A British Army doctor, writing from France to a former colleague, complained that the 'aspirin' he was doling out to the troops in his section was 'very chalky and difficult to swallow. Its qualities are more emetic than analgesic and the men are most reluctant to take it.'

At least the army was getting some. The *Prescriber*, the same journal that in March 1915 had boasted about the imminent availability of British aspirin, was complaining two months later that it had failed to materialize. 'All we have heard of the British product is a statement by the chairman of one of the largest hospitals that they had received the first consignment of 56lbs of sodium salicylate – the mountain conceived and brought forth a mouse.' And sodium salicylate was, of course, just one of aspirin's poor relations and nowhere near as effective as the real thing.

Even though Bayer's British subsidiary had been closed down along with other German firms at the start of the war, some of these problems would eventually have come to the attention of the Bayer board; throughout the conflict, businessmen on both sides were remarkably adept at keeping in touch with what was happening to their former interests. But whatever patriotic satisfaction Carl Duisberg and his colleagues may have felt on hearing the news, it would have been offset by the long-term implications. Anyone in Britain could now theoretically produce and sell aspirin, and the longer the war went on, the greater was the possibility that someone there would find a way to make a commercial product that rivalled Bayer's in quality. Once that happened, it would be well nigh impossible for Bayer to reclaim the drug as its own – unless Germany won the war of course.

In the event, though, the most damaging blow to Bayer's once proud monopoly in Britain and her dominions came not from the UK at all, but from a small high-street pharmacy on the other side of the world.

George Nicholas patted out the last of the flames on his young assistant's coat and looked around at what else needed to be done. The room was filling with fumes and smoke and a small tin of ether still blazed merrily away on the bench alongside him. Immediate action was called for but he couldn't see anything to douse it with. With a sinking heart, he realized he had only one option. He knocked the tin to the floor and sat on it. As the flames died beneath him and the first unpleasant whiff of scorched flannel reached his nose, he wondered for the hundredth time what had gone wrong. There had to be an easier way to make aspirin.

George Richard Nicholas had been through many such moments in the preceding weeks. He'd filled his chemist's shop with noxious fumes and had twice almost blown the room apart. He'd worn himself out in the process, losing a great deal of weight and once, albeit temporarily, even going blind. But Nicholas was a man with a mission. There in his little shop, in a quiet Melbourne suburb, he was determined to produce the drug that had taxed some of the best brains in Europe.

Until 1914, Australia, like the rest of the world, had bought most of its aspirin from Bayer in Germany. But when Australia had loyally followed Britain into a declaration of war those supplies had dried up. Given the problems the mother country was experiencing in meeting its own demand for ASA, there seemed little prospect of getting any more from Britain for some time to come, or indeed any of the other vital chemicals the country needed. Australia would have to make her own. So, shortly after war broke out, Australia's Attorney General, Billy Hughes, announced that German chemical patents and trademarks would be suspended and awarded to any domestic producer who could meet the appropriate quality standards. With a surfeit of exciting war news coming from Europe, his statement didn't attract much attention at the time, but thirty-one-year-old George Nicholas read it and was inspired.

His problem was that although he was a qualified pharmaceutical chemist, the day-to-day reality of his life was serving the customers in the Junction Pharmacy. Turning the premises into a proper laboratory just wasn't that simple and it had been some years since his student days when he'd learnt how to compound and distil complex chemical substances.

In theory, of course, aspirin was easy enough to make. You heated salicylic acid powder with acetic anhydride and then condensed the combined vapours back into a liquid. When this dried, the white crystalline powder that was left should be acetylsalicylic acid. The more care you took over mixing, heating and then cooling the chemicals, the purer the resultant substance should be. The aim was to reduce as much as possible any free salicylic acid left in the final compound, as this was the cause of the nasty gastric side effects that had traditionally been the drug's greatest problem.

But Nicholas's first attempts ended in abject failure. He couldn't afford a vital piece of equipment called a reflux conductor, through which the all-important vapours could be cooled. So he had to improvise, using bits and pieces of kit that he had lying around in his shop and a few pots and pans loaned to him by his wife. Eventually, after experimenting explosively with different mixtures and heating times, he obtained acetylsalicylic acid of a sort, but the pinkish soggy mess was a far cry from the pure white crystals of Bayer aspirin that he'd been selling to the public before the war.

Salvation arrived in the shape of Harry Woolf Shmith, an amateur inventor and freelance industrial chemist. He came into the shop to buy something one day, got chatting to Nicholas about his efforts and offered to help. With two of them focused on the task things went much faster, particularly when they hit upon the idea of dissolving the pinkish residue of their experiments in ether and then recrystallizing it. The new procedure took a while to work out but after a few weeks they were able to gaze down at some pure white crystals nestling in the bottom of a Petri dish and declare themselves satisfied. They had made aspirin.

The next step was to alert the Australian government to their achievement. But that proved difficult. Letter after letter to Canberra went unanswered. There was a war on, everyone was busy, and who, after all, had ever heard of George Nicholas? Finally, though, the matter came to the attention of Attorney General Billy Hughes. He was persuaded to authorize a federal government analyst to carry out some tests and even attended some of the sessions himself. The results were announced on the front page of the *Melbourne Herald* on 17 September 1915, tucked in alongside a trailer for a forthcoming special edition on Australian casualties at Gallipoli. With the ghastly consequences of the war so evidently in people's minds, the patriotic politician couldn't resist taking a swipe at his nation's enemies.

AUSTRALIAN ASPIRIN
IS GRANTED LICENCE

'Purer than German,' Asserts Mr W. M. Hughes

A Victorian Triumph

Mr W. M. Hughes, the Federal Attorney-General, announced today that after a test in his presence, he was satisfied that a sample of aspirin made in Australia was purer than the German product, which practically held the market for the drug, and that he had accordingly granted a licence to the makers, Messrs H. W. Shmith and G. R. R. Nicholas, to manufacture and sell aspirin in the Commonwealth, despite the fact that aspirin was a German trade mark.

'The conditions of the licence,' said Mr Hughes, 'ensure that the drug shall comply strictly with the requirements of the British Pharmacopoeia, and that the conditions of manufacture and price at which it shall be vended shall be satisfactory to the Attorney-General.'

A test was carried out on Tuesday by the Federal Government analyst, in the presence of the Attorney-General of preparations, including those of Bayer

> and Company of Germany, and of Messrs Shmith and Nicholas.
>
> 'Of these,' said Mr Hughes, 'only that prepared by Messrs Shmith and Nicholas was found to be absolutely pure. In the circumstances, therefore, the Australian medical profession has the positive assurance of the Australian-made drug, that it contains no free salicylic acid and that it in all respects complies with the requirements of the British Pharmacopoeia. The Australian public may rest confident that while encouraging their fellow-citizens and their enterprise, they are getting an absolutely pure and reliable article.'
>
> Mr Hughes added that he desired to call public attention to the fact that while it is not proposed to permit further importations of German-made aspirin, yet considerable quantities are in the drug stores throughout Australia, and the public should bear in mind that none of this absolutely complies with the requirements of the British Pharmacopoeia – that was to say that they contained free salicylic acid. Therefore the public should refrain from purchasing the German drug both from patriotic and prudent motives.
>
> Messrs Shmith and Nicholas state that they intend to vend the Australian made drug under the trade name of 'aspirin'.

Putting to one side Billy Hughes's loyal but defamatory comments about Bayer aspirin, the implication any average reader would have taken away from this article was that a leading Australian pharmaceutical firm had managed to beat the enemy at their own game. There wasn't a word about the fact that the manufacturers were actually working from a single-storey chemist's shop just down the street. And therein lay the new enterprise's next big challenge. They had managed to produce acetylsalicylic acid in their primitive laboratory – although any number of people in the world had done that. They had obtained a remarkable endorsement from one of the country's leading politicians, in effect an instruction to all good citizens to buy this nice new Australian product – and that was certainly worth having. But now they actually had to make the drug, and in commercial quantities too.

With almost no money or equipment between them, Nicholas and Shmith clearly weren't going to get very far on their own and so they began looking for help. Their first recruits were from their respective families. George signed up his elder brother Alfred, a Melbourne businessman whose import trade was being hit by the war, and he in turn recruited J. Wilhelm Broady, one of his old partners. Harry Shmith brought his father into the firm.

As it happened, this did little to relieve the fledgling company's chronic lack of capital, as none of the new partners had much cash. But somehow they persuaded suppliers of raw materials to extend them lines of credit and they borrowed an old hand-cranked tablet-making machine. Then they cleared out the basement of Alfred's old office to use as a packaging centre, called in their respective wives and a few local girls to help, and slowly and with many hiccups went into production.

The first few months were tough. In October and November 1915, they barely sold enough of the new aspirin to cover their costs. To make matters worse, just as sales began to pick up, an importer of British medicines, waiting to receive a rare shipment of ASA from the UK (and presumably in an attempt to see off these new rivals) spread the rumour in the press that Shmith–Nicholas was a front for Bayer and hinted that Billy Hughes had been bribed to grant them a licence. It was nonsense, of course, but in the hysterical anti-German atmosphere of the times it was a clever and damaging slur, given more credence by the Teutonic-sounding names of the Shmiths and J. Wilhelm Broady. Although Alfred Nicholas fought back, declaring indignantly that his family could trace its lineage back to Cornish mining stock and insisting that the government examine everyone else's birth certificates, the furore proved too much for the Shmiths and Broady. They sold their stakes in the business back to the Nicholas brothers and left them to struggle on alone.

Things stabilized for a while, helped by further disruption to shipments of ASA from Britain,* and sales picked up again. By the end of 1916, the company was selling around £1,300-worth of aspirin a month and the future looked more secure. But then in early 1917 a fiercely anti-German MP called W. H. Kelly raised the Bayer rumours once more, this time whipping up a storm in Australia's parliament that the government couldn't ignore. The Board of Trade was forced to renounce the Nicholas brothers' exclusive right to call their product aspirin.

George and Alfred had to think of a new name fast, and put the most positive spin on the news that they could. On 21 May 1917, they placed an advertisement in the *Australasian Journal of Pharmacy*:

---

* A large shipload was torpedoed off the coast of France in December 1916 while en route to Australia.

SHOULD ASPIRIN BE RE-NAMED?

Important Announcement by the Proprietors of Nicholas-Aspirin.

'Now that Aspirin is selling so well, it would surely be a mistake on your part to change the name.' Such is the tenor of various letters we have received from all parts of Australia, and the matter is so important – both to the trade and ourselves – that we should like to discuss it fully.

In view of the national sentiment in regard to German Trade Names, the manufacturers of Nicholas-Aspirin feel that the time has arrived when a change should be made.

We have, therefore, taken the first step by registering the name Nicholas-Aspro, which is now our property. It is our intention to substitute this name for Nicholas-Aspirin. Pharmacists of Australia can rest assured that the purity of aspirin under the new name ASPRO will be zealously guarded under the Federal Commonwealth Government supervision as before. In other words nothing has been changed but the name . . .

Years later the company would claim that the origins of its aspirin's new identity lay somewhere in the last two letters of Nicholas and the first three of its product, but the likely truth is that in those difficult times Aspro was the closest to aspirin the brothers felt they could get away with. The move was not without its risks, however. Rebranding a product barely two years from its launch could easily backfire. What if the public didn't take to it? How could they market a drug that no one had ever heard of? Then, by one of those strange quirks of fate that so often dictate the course of events, someone turned up with exactly the right set of skills they needed – Herman George Tankersley Davies.

He blew into their lives on a wet and windswept autumn evening in 1917. Alfred was sitting in the office, running his eyes over yet another depressing set of sales figures, when a damp and tousled head appeared around the door.

'G'day, mate. Any business?'

George Davies, former owner of a failed clothing business (it had produced cheap suits from redyed army surplus khaki) and now touting for orders on behalf of a local printing firm, had come to say hello. A huge, scruffy New Zealander, all chat and charm, he stood there dripping water on to the threadbare carpet and, like any good salesman, got Alfred talking. This was no mean feat. Unlike his more ebullient younger brother, Alfred could be fussy and dour, prone to snappiness when things weren't going right. But Davies was an old pro with reluctant customers and before long he had wheedled the whole Aspro saga out of him.

It was then, Davies recalled later, that he saw the light. He had stumbled on an opportunity too good to be missed, a once-in-a-lifetime chance. As George Nicholas came in to join his brother, Davies started his pitch, striding the room, throwing out ideas. They were sitting on a gold mine, he told them. But they weren't promoting it properly. What they needed was someone with imagination, drive and initiative. Handled right, Aspro could make them millions. And he was just the man who could do it for them.

The following day, carried away by his optimism, the Nicholas brothers signed him up on a retainer of £4 a week and a 1 per cent commission on all the company's sales. It was probably the best business decision they ever made. Just as he promised, George Davies was going to make them rich.

The consequences of these events for Bayer would be dire but for the moment they were unfolding a long way from its area of main concern – the United States of America. When war broke out in Europe, Germany was still at peace with the United States and Bayer still had three years left on its vital American aspirin patent. Three years in which to secure the product's future in the richest market in the world. It was hard to think of anything else.

For the United States, the most immediate impact of hostilities in Europe was the interruption of trade. Britain was determined both to deny Germany and its Allies access to vital war supplies and to disrupt their economies. Prior to the war, Britain had been a party to international agreements aimed at allowing sea trade between non-belligerents to continue in the event of any conflict. Germany and the United States were at peace and legally therefore their trade should have been allowed to carry on uninterrupted. But Britain rapidly reversed out of this policy when it became clear that the war was going to last longer than expected. Although maintaining her own ability to trade with the great neutral would be essential, it was equally important to cut the Central Powers off from the same nourishment. The Royal Navy's blockade was extended into the Atlantic.

This infuriated not only Germany and Austria (who would no doubt have done the same thing had they been able) but the United States as well. President Woodrow Wilson believed, as did most of his fellow citizens at the beginning of the war, that America's interests were best served by staying well clear of any foreign adventure. Neutrality would enable the United States to play the part of impartial mediator, the New World saving the Old World from its follies. In the meantime America would do business with whomsoever she liked – and as US trade with the Central Powers was worth $169 million in 1914 it was not something to be discarded lightly.

But the practical reality was that by 1915 the Royal Navy controlled the Atlantic and was not only attacking German merchant ships on their way to America, but was also beginning to stop and search German-bound US vessels, either impounding their cargoes as embargoed contraband or turning them home. This raised howls of protest from the State Department and for a time diplomatic relations between Washington and London became positively frosty. Indeed, things might have become even more serious had not Britain and France's own trade with America begun to take up some of the slack. Fortunately for the Allies, as this increased it began to dull the edge of Washington's fury, but the situation remained difficult for many months to come.

However, the disruption of business between Germany and the United States had other consequences, one of the most significant being that America could no longer rely on imports of hundreds of important commodities. Chemicals, particularly coal-tar products, were at the top of the list. The American synthetic chemical industry, prior to the First World War, was a mere infant compared to its German counterpart, with neither the knowledge nor the infrastructure to compete. And the big German chemical companies had done their best to keep it that way. Only a few coal-tar medicinals, for instance, had been actually manufactured in the United States. Bayer made acetylsalicylic acid at Rensselaer, and at other smaller plants companies like Heyden and Fries Brothers produced some salicylates, but there wasn't much else. Other chemicals were even more scarce, vital intermediates that were used in a huge array of chemical processes and products from dye products to pharmaceuticals.

German sympathizers in the United States used these shortages to propagandize their own cause. In July 1916, they had a notable coup to celebrate when the German submarine *Deutschland* surfaced dramatically off Baltimore with a cargo including 300 tons of concentrated dyestuffs (around 1,300 tons in real terms). Outwitting the British in such an exciting way caught the American public's attention and was seized upon by pro-German newspapers as evidence of the futility of the Royal Navy's blockade. But although such romantic ventures made the headlines (the *Deutschland* returned with another cargo of drugs and chemicals four months later), they did little to solve America's pressing needs.

Bayer, ironically, was one of those most affected.

At first, its business in the United States had been relatively untouched by events in Europe. Hedging against the possibility that American neutrality would not last for ever and that one day Bayer's operations there might conceivably be vulnerable to anti-German sentiment, Carl Duisberg had sanctioned a discreet corporate reshuffle just before the war. Ownership of Bayer's US assets and trademarks (including that of aspirin) was transferred to the company's American subsidiary, while its patents were given to a specially created firm called the Synthetic Patents Company. The hope was that if a conflict broke out, the moves would disguise Bayer's German origins and enable it to carry on trading. But in reality, of course, Germans employed by Bayer held the stock and there was little day-to-day change in the way things were run. The company's principal obsessions continued to be the looming expiration of its US aspirin patent, the difficulties caused by the restrictive American Medical Association's guidelines, and the best ways to combat the pharmacists and counterfeiters who sold fake or smuggled versions of the drug.

Eventually, though, the war gradually began to influence events in a way that no one had foreseen. The blockade made it increasingly difficult to maintain communications between New York and Bayer headquarters in Germany, and Carl Duisberg was forced to allow his subordinates in America a degree of latitude that would have been unthinkable a few years before. This breakdown in central control would prove disastrous because it led directly to one of the most damaging episodes in the company's history – a bizarre conspiracy known as the Great Phenol Plot.

Phenol, as struggling British pharmaceutical companies had found to their cost, was vital to the manufacture of salicylic acid but also to the production of the high explosives known as picric acid and trinitrophenol (a kind of TNT). It was therefore a commodity of the utmost strategic importance that the British authorities were determined should be strictly controlled. None of it, from Germany or indeed from Britain, was getting across the Atlantic. As stocks in America ran low (domestic producers couldn't yet make enough of it on their own), prices rose sharply. Manufacturers of salicylic acid were unable to meet the demands of Bayer's Rensselaer plant, which made all its American aspirin. By April 1915 the situation became so serious that the factory was almost shut down. To make matters worse, when rumours spread that aspirin might no longer be available, counterfeiters stepped in to fill the gap. For a company that had spent years trying to persuade and bully pharmacists into stocking only the genuine article, this was a disaster. In desperation, and cut off from Carl Duisberg's advice, Bayer & Company's executives sought help elsewhere. They turned to Hugo Schweitzer.

Schweitzer had emigrated to America in 1889 after gaining a doctorate in coal-tar chemistry from the University of Freiburg and soon found work with Bayer's American subsidiary. After a few years he was made head of its pharmaceutical department in New York and was instrumental in persuading Carl Duisberg of the wisdom of opening the Rensselaer plant. Although he didn't especially like America, he had found it convenient to become a US citizen and did well as a consequence, moving out of the company's direct employ to become a highly paid consultant and a leading figure in the expatriate German community. Duisberg was particularly impressed by him and even asked him to become President of the Synthetic Patents Company that Bayer had set up just before the war.*

---

* Such was Schweitzer's influence in the chemical community that he was chosen in October 1906 to give the keynote speech at a remarkable event – a dinner held in honour of William Perkin, the discoverer of mauveine dye. The sixty-eight-year-old Englishman, then in the twilight of his career, had been persuaded to cross the Atlantic to meet his admirers, and no effort had been spared to make him welcome. Many of those who attended the glittering celebration at Delmonico's restaurant realized that Perkin's invention, fifty years earlier, had been the starting point of the world's pharmaceutical revolution. But it was Hugo Schweitzer who put those thoughts into words. 'It is hard to imagine today what an epoch-making idea it was,' he said. 'It was truly the spark of genius.'

When the conflict broke out, Schweitzer became a very public advocate on Germany's behalf. American public opinion was, if anything, slightly pro-German at the start of the war, and he did all he could to try to keep it that way. He delivered anti-British tirades at mass public meetings, founded the German Publication Society to disseminate translations of Germany's greatest writings, and even tried to buy the *New York Evening Mail* as a vehicle for propaganda. When a German U-boat torpedoed the British ocean liner *Lusitania* in May 1915 and 1,000 people, including many Americans, died (an act which began to swing US public opinion against Germany, and, incidentally, hushed many of the State Department's complaints about the British blockade), Schweitzer robustly defended the act by saying that anyone who sailed on an English ship 'committed suicide outright'.

Some of his boldest patriotic statements came in his writing and speeches about the mighty German chemical industry. In one typical pamphlet he said:

> In no other field has German efficiency proven its superiority as in that of chemistry. While this was anticipated before the war, it is no exaggeration to state that the German chemist has so far contributed as much, if not more, to the success of the campaign than the strategist, the army and the navy, and that therefore, the present holocaust may justly be called the 'chemists' war'.

But these very high-profile activities masked his other more clandestine role – as a spy. Recruited by the German Secret Service before the war and given the code number 963192637, his principal task was to act as a conduit between the German Embassy and one of their most important agents.

The German Ambassador at the time, Count Johann Heinrich von Bernstorff, not only had the diplomatic role of trying to keep America neutral, he also had the more delicate tasks of frustrating US exports of armaments to his country's enemies and obtaining the same products for his own side. His principal aide in this was Heinrich Albert, a German Interior Ministry official seconded to the Embassy for the duration of the war. But as both von Bernstorff and Albert were kept under surveillance by the American authorities (and presumably by the British too) they had to have trusted middlemen who could deal on their behalf with their network of agents and sympathizers. Hugo Schweitzer was their cut-out for Walter Scheele, a chemist at the New Jersey Chemical Company and one of Germany's most important industrial agents in America in the early years of the war. A highly experienced scientist in his own right, Scheele was reputedly part of a team that invented mustard gas in 1913, and passed the formula secretly through Schweitzer back to Germany (where it was produced at Bayer's Leverkusen plant). Under Schweitzer's direction he also prepared incendiary devices that were planted in British vessels anchored in New York harbour and worked with him to devise a method of disguising American oil as fertilizer so that it could be smuggled past US Customs to Germany.

It was exactly this kind of resourcefulness that brought Bayer to Schweitzer's door with a plea that he help them find a fresh supply of phenol. True to form, he began making discreet enquiries and soon devised a solution.

Bayer wasn't the only business suffering from the phenol shortage. Another was that of Thomas Edison, the famous inventor, who needed it to make phonograph records, one of his most successful and popular creations. After trying, and failing, to buy any phenol on the spot market, Edison had decided with his customary élan to make some himself. He worked out a way to synthesize it from benzene and in June 1915, the newspapers carried a story that he was building a refining plant in New Jersey to produce it. The factory would produce around twelve tons of phenol a day, of which three tons would be surplus to Edison's needs. The excess would be sold through his agent, the American Oil and Supply Company, which was now open to offers from interested parties.

Two days later American Oil announced that with Edison's agreement it had sold the phenol on contract to an organization called the Chemical Exchange Association. This was a firm previously unknown to the sellers and so they had asked for – and secured – an advance cash deposit of £100,000. The overall price, thought to be very high, was not specified. Speculation was rife among other potential purchasers, including several armaments exporters, as to who might be behind this new firm. But no one seemed to know.

In actual fact it was a creation of Hugh Schweitzer. Immediately on hearing of Edison's plans he had arranged a meeting with George Simon, the general manager of the American branch of Chemische Fabrik Von Heyden, and Bayer's principal supplier of salicylic acid. Together they worked out a deal. With Schweitzer supplying the money (which actually came secretly from Heinrich Albert's espionage funds at the German Embassy), Simon would discreetly buy the phenol through the bogus Chemical Exchange Association. Some of it would be used to meet the contract with Bayer's aspirin plant at Rensselaer; the rest would be sold back to Schweitzer at cost for disposal elsewhere.

A few weeks later Schweitzer gave a lavish private dinner at New York's swanky Hotel Astor in honour of Heinrich Albert. It was a happy evening because behind all the backslapping, champagne and cigars lay the knowledge that Schweitzer had pulled off a remarkable coup. George Simon's salicylic acid plant was now back in business and Bayer Aspirin was pouring off the production lines once more. Schweitzer now controlled one of the few available sources of phenol in America and was set to make a fortune. Furthermore, all this had been paid for out of Germany's secret funds, which Albert had been happy to sign off because it meant that phenol equivalent to four and a half million pounds of explosives wasn't going to end up in Allied munitions factories.

The jubilation was short-lived. On 24 July, Heinrich Albert mistakenly left his briefcase on a train into Manhattan. When he returned in a panic to try to find it, fellow passengers told him a young man had taken it. Unfortunately, the man turned out to be an agent of the US Secret Service who had been tailing Albert for weeks, and the bag was stuffed full of secret papers. They contained information about networks of unidentified German sympathizers, coded references to sabotage, and documents about the recent phenol purchases.

The papers were not detailed enough to give the American authorities sufficient grounds for arrests, so they leaked them to the press instead. On 15 August, the *New York World* splashed the story on its front page and accused Albert, Hugo Schweitzer and Count von Bernstorff of subversion. The three men, claimed the paper, had been trying to undermine Anglo-American interests for years through sabotage and propaganda and by 'stealing' vital American chemicals like phenol.

Over the next few days Albert, von Bernstorff and Schweitzer were besieged by a hungry press pack demanding a response. The three men did their best to shrug off the attention, Schweitzer at one point claiming that all the phenol was to be used for making hospital disinfectant, while Albert and von Bernstorff sought refuge behind the walls of the German Embassy. Eventually the fuss died down, but the damage was done. A short while later an embarrassed Thomas Edison decided to sell the rest of his phenol to the US military (although by then enough had been secured to keep the Bayer plant going). Hugo Schweitzer's credibility, meanwhile, was shot. Although he carried on with his propagandizing, he knew he was now a marked man, under continuous surveillance that severely limited his effectiveness as an agent. Eighteen months later he contracted pneumonia and died. When the New York police searched his apartment they found secret code books, copies of his flamboyant anti-British speeches and notes on the various ways of making aspirin.

Bayer was badly tainted by the scandal. Its attempts to distance itself from the political machinations of the German Embassy convinced no one and from then on the suspicious US authorities would monitor the company's every move. In the process it squandered whatever influence it might otherwise have had in Washington and probably sealed the long-term fate of its most precious asset as well. Its executives didn't know it, but their hold on aspirin was slipping away.

If that wasn't yet apparent it was because all the company's attention was bent on squeezing the last drop of gravy out of its US patent. With only a year left to run and with rival American chemical companies straining at the leash for the moment when they too could jump on the aspirin bandwagon, Bayer now decided to press ahead with a marketing strategy that would outflank its carping critics in the American medical establishment. The time had come to start selling aspirin direct to its users. The drug's days as an ethical medicine were numbered.

In June 1916, *Printer's Ink*, a trade journal for the advertising industry, told its readers that Bayer was launching a discreet newspaper advertising campaign to familiarize Americans with its aspirin trademark.

> The reader will not be urged to buy anything. The product will not be suggested as a remedy for any ailment. The uses to which it can be put will not be mentioned. The sole object of the publicity is trademark identification.

When the ads appeared a few weeks later, they could barely have been more reticent. Under the single headline 'Bayer' a picture of the company's aspirin box was followed by some brief words of copy: 'Tablets of Aspirin. The Bayer Cross on every package and tablet of Genuine Aspirin protects you against all counterfeits and substitutes.'

But this false modesty did little to save Bayer from the wrath of the AMA. Its response was swift and furious. Its journal reminded doctors that for seventeen years it had been legally impossible in America for anybody other than Bayer to sell acetylsalicylic acid and that the public as a consequence 'have been made to pay exorbitantly for the monopoly our patent office granted this firm'. The moment the patent was up, doctors should prescribe alternatives.

It may seem strange now that such an innocuous set of advertisements should have provoked such ire, but the fact is that by the standards of the time Bayer was considered to have contemptuously flouted the ethics of the profession. It was as though all the AMA's long campaigns against branded and patent remedies counted for nothing. Once this particular genie was let out of the bottle, what would stop other medicine makers doing the same thing? And then how would the public distinguish between a legitimate remedy and a fake? Immediate action was called for. Bayer aspirin was dropped from the AMA's official list of recommended medicines.

But Bayer seemed almost past caring what the doctors thought. With one eye on the calendar and the fast approaching day when its patent would be lost, and another on its sales which stayed buoyant despite the AMA's protestations, the company knew that it had to cash in on the drug's popularity while it was still its exclusive property. In January 1917, it began advertising again, this time in an attempt to forestall what it saw as its greatest threat – that rivals would try to steal its brand name once the patent expired the following month. It began with a full-page ad in the AMA's own journal (*shown overleaf*):

Shortly afterwards the company advertised again, this time to tell the pharmaceutical community that it had put fire into its words. It had filed suit against the United Drug Company of Boston for trademark infringement. The rest of the industry should take note.

But Bayer's warning was out of sync with the tenor of the times. For years its US rivals had licked their lips at the profits Bayer was making. In the three years since 1914, American patients had consumed almost two million lb of aspirin with a retail value of $25 million. That was a vast sum of money by the standards of the day and the fact that most

'ASPIRIN'

Trade-Mark

The Trade-Mark 'Aspirin' (Registered US Patent Office) is entirely separate from the patent on Acetyl Salicylic Acid and will not expire with this patent. The Trade Mark 'Aspirin' remains our exclusive property, and therefore only acetyl salicylic acid manufactured by the Bayer Co., Inc., can be marketed or sold as 'Aspirin'.

Any violation of our trade-mark rights will be vigorously prosecuted.

of it had gone back to a country with which the United States might soon be at war – a country which many thought had long maintained a worldwide monopoly in synthetic chemicals for its own sinister purposes – gave their resentment an edge. Several companies, including Dow Chemicals and Monsanto, had set up production lines to make ASA as soon as the patent expired and now they made plain their determination to call it by the name that all its consumers had got used to.

Events were on their side. Germany, encouraged by a belief that the Allies were losing the will to prosecute the war (and by the beginnings of a revolution in Russia) had started an unrestrained U-boat campaign in the Atlantic in an attempt to put extra pressure on Britain and France. Unfortunately, it sank several American ships in the process. It had also unwisely become involved in a plot to persuade Mexico to attack the United States. As news of these events was reported in America, so the mood darkened. It finally became plain to Bayer's New York-based executives that they might soon be enemy citizens running an enemy business. A scramble to complete the process of disguising the company's US holdings in shell companies ensued – and continued even after the United States' declaration of war on Germany on 6 April 1917. But it was all too late. Nemesis appeared in the shape of A. Mitchell Palmer, the Alien Property Custodian (APC).

The office of the APC was a consequence of the Trading with the

Enemy Act, which created a body to take over all enemy property and hold it in trust until the end of the war. A. Mitchell Palmer, an arrogant and dogmatic former congressman from Pennsylvania, was its first incumbent. To him fell the task of seizing Germany's sizeable assets in America, worth around $950 million, many of which had been squirrelled away in dummy businesses. To help him unravel this corporate spaghetti, he set up an aggressive enquiry arm, run by a former New York Assistant District Attorney called Francis P. Garvan. This agency, the Bureau of Investigation, was the forerunner of today's FBI.

Together, A. Mitchell Palmer and Francis Garvan went to work and Bayer, with its connections to the notorious Hugo Schweitzer still fresh in the authorities' minds, was a natural target. As Garvan, a fanatical anti-German, began delving into the company's confusing corporate structure, Mitchell Palmer announced he was seizing all Bayer's assets, including its patents and trademarks, and appointing Americans to its board for the duration of the war. One appalling question was uppermost in the minds of the few German executives left clinging on to their corporate posts: would the Americans ever give the company back?

The years before the war had seen Farbenfabriken vormals Friedrich Bayer & Company complete its transformation from a middle-sized dye and coal-tar derivatives company into one of the biggest chemical firms in Germany. Boosted by the profits from aspirin, it had continued to invest heavily in developing new and ever more complex compounds for use in a myriad of industries at home and abroad. The pharmaceutical division was among the most active, devising antiseptics, barbiturates (a new class of sedative and hypnotic drugs), heart treatments, leprosy treatments, and much more. Some enjoyed a brief flurry of success before being superseded by more effective treatments developed by rivals. Others, like the famous Salvarsan syphilis drug that would restore the health, if not the reputations, of so many young soldiers in the First World War, proved hugely profitable. But none of them outshone aspirin, which was in a class of its own. Every young Bayer pharmacist must have dreamt of emulating that particular discovery, even though wiser heads in the laboratory would have reminded them gently that fate, fickle fashion and advancing medical knowledge could bring down even the most successful products in time. Look at heroin, they would have said. Once trumpeted by its Bayer patron, Heinrich Dreser, as a safe and effective cough medicine, its true habit-forming nature had eventually become apparent. In 1913, there was a rash of bad publicity when heroin-related admissions mushroomed in hospitals across the east coast of America. Reluctantly, Bayer decided to stop making the drug and within five years it had been declared an illegal substance in much of the world.*

Carl Duisberg had watched over these developments, good and bad, with all the attention to detail and single-minded determination of an obsessive parent. He hated it when things went wrong and did everything he could to put the company in an unassailable position, impervious to shocks. One of his most enduring achievements had been to drive through the construction of the massive Bayer plant at Leverkusen, near Cologne. This huge works, built on a twenty-five-acre site along the banks of the Rhine, was completed just before the war and was heralded as one of the most remarkable manufacturing facilities in the world. Raw materials came in from riverside wharfs and left as finished product by rail. Its grid-pattern streets and interconnected factories made it a model for other companies to aspire to – efficient, cost-effective and vastly impressive. And every one of the many thousands of workers who streamed in and out of its gates would have been constantly aware of Duisberg's domineering presence. He had moved the company's headquarters from Elberfeld to Leverkusen, and built a neo-classical Great Hall as a centrepiece of its operations. He lived there too, in imperial splendour, in a house surrounded by fountains and formal gardens, backing on to the buildings where the all-important dyes, chemicals and drugs were made.

The flamboyance had little to do with gilding the vanity of the great man. It was a statement that Bayer was a force to be reckoned with –

---

* In New York, some of these early heroin addicts fought over scrap metal they collected on the streets to pay for their habit – in the process earning themselves the nickname 'junkies'. Even though it was declared illegal, the drug remained the narcotic of choice for most of America's addicts, particularly when the Mafia became involved in its trafficking in the 1930s. Heinrich Dreser left Bayer in 1914 but invested some of his massive heroin (and aspirin) royalties in a new pharmacological institute in Düsseldorf. Following his death from a cerebral haemorrhage ten years later, there were rumours that in his last years he had become a heroin addict himself.

an industrial giant of the twentieth century. But Carl Duisberg was a realist as well as a visionary. He knew that the company, huge though it had become, was only one of a number of such giants in Germany, all in ferocious competition with one another. And because such corporate bloodletting was harmful, wasteful and expensive, he decided to try to prevent it.

On his trip to the United States in 1903 he had learnt much about how American industries conducted their business, and in particular the way that cartels, like that led by John D. Rockefeller's Standard Oil, managed to smooth out aggressive competition by agreeing to co-operate on pricing and supply. Back in Germany, he set about trying to convince his peers in the chemical industry that this was a model they ought to adopt. He was only partially successful. There were powerful egos involved and the men who owned and ran Bayer's rivals regarded his grand ambitions with suspicion. None the less, in 1904 he sat down with the bosses of two of the largest – BASF and Aktiengesellschaft für Anilinfabrikation (the precursor of today's Agfa company), and agreed a loose confederation, known as the Dreibund, or Triple Association. At the same time, two of the other largest German chemical firms, Hoechst and Leopold Cassella & Company, formed their own combine. It wasn't ideal as far as Duisberg was concerned, as he would have liked to have all these firms under one umbrella. But he was patient, confident that one day commercial logic would drive them all closer together. In the meantime, the two parallel associations prospered, avoiding destructive competition whenever possible and exerting a stranglehold over markets at home and abroad.

The war gave Duisberg the spur he was looking for. As the British blockade set in, the huge monopolies these companies had enjoyed over the world's synthetic chemical supply began to dissipate. Fortunately, as export markets dried up they were replaced with orders from the imperial war machine. The German Army's appetite for munitions was as vast as its enemy's, but unlike the Allies the Germans found it difficult to get supplies from overseas. Bayer and the other companies had to fill the gap. Factories that had previously made fertilizer now made high explosives. Scientists who in peacetime had worked on pharmaceuticals or dyes now turned their hand to poison gas, first deployed at Ypres in April 1915. As Hugo Schweitzer had written in one of his pamphlets, it had become a 'chemists' war'.*

But inevitably, the more the chemical giants collaborated on providing the Kaiser's army with weapons, the closer they came together. Duisberg realized that if these relationships could be cemented now, he could eliminate post-war rivalry and ensure the continued growth and prosperity of all concerned. The strength of his argument was persuasive and in January 1916 the two German chemical federations joined into one loose combine. Each of its constituent companies would continue to be autonomous and would retain its corporate identity (Bayer would still be Bayer, Hoechst would still be Hoechst), but they agreed to consult each other on research, production and sales, and to share profits on a commonly agreed scale. It would be several years before the full logic of these deals was realized through a proper merger, but even in this half-formed state it was still the largest industrial group in the world. Carl Duisberg became the chairman of this colossus, which was called the Interessengemeinschaft der deutschen Teefarbenfabriken – the Community of Interests of the German Tar-Dye Factories. Everyone quickly abbreviated this to a name that would one day resonate for all the wrong reasons: IG Farben.

It was ironic that even as Duisberg began to realize these long-held dreams of amalgamation, one of the key parts of his own empire was falling away. Aspirin was the most precious of all Bayer's assets and Duisberg had tended it like a fragile flower ever since its first appearance. That care and attention had been repaid a thousand times over as the drug had become one of the most popular and profitable in the world, but now in his moment of triumph he was beginning to lose control of it. A true patriot, he had few misgivings about the war when it began, but the longer it went on, the more he must have grieved at the sight of aspirin falling into enemy hands. The loss of the British market was a bitter blow, but at least it came in the heady days of the conflict when there still seemed every possibility that it would be

* Duisberg was enthusiastically involved in the development of at least one poison gas – phosgene – and even tried it out on himself. On 3 March 1915 he wrote to Major Max Bauer, the Supreme Command's liaison officer to the German chemical industry: 'How uncomfortable it works you may best gather from the fact that for eight days I have been confined to bed, although I inhaled this horrible stuff only a few times . . . if one treats the enemy for hours at a time with this poisonous gas-forming product, then according to my view he will not immediately leave the country.'

regained. The news from Australia had been a gnat bite in comparison and it would be some time before it became apparent just how large a wound in Bayer's side it would turn out to be. But the developments in America caused Duisberg real concern. Effectively cut off from his managers there, he'd had to watch helplessly from afar as they blundered in and out of scandal and twisted and turned under the scrutiny of the American Medical Association. He had always been a little wary about the correctness of advertising as he was a firm believer in the principle of ethical medicines – or at least in keeping the doctors on side – and as long as the profits kept coming in he was reluctant to push things too far. But the imminent loss of the US patent had driven everybody to take risks that Duisberg might have drawn back from had he been personally in charge. And then to cap it all, in April 1917, the United States had declared war and taken control of all Bayer's American assets. Sitting in the Great Hall at Leverkusen, unable to influence events, Carl Duisberg must have been a very frustrated man.

Perhaps he believed for a while that there was still a chance that solutions might be found, that events might unfold the right way. Germany was still in the war, and although acute food and fuel shortages were beginning to bite at home, her armies were still in the field. In many ways, it was possible to hope that the tide was turning in Germany's favour. The latest U-boat offensive on Allied shipping was proving effective and at least one of Germany's enemies, Russia, had succumbed to its own internal chaos and taken itself out of the war.*

If so, those hopes were short-lived. As Germany's last offensives expired amidst the mud and slaughter of the Marne in the summer of

---

* What would Duisberg have thought had he ever discovered the part that aspirin played in that Russian collapse? Alexei, the son and heir of the last Czar, suffered from haemophilia and the dreadful pain and inflammation caused by blood flowing into his joints. It's very likely that to ease the pain the imperial doctors prescribed what, even then, was the world's most popular analgesic, not knowing that aspirin actually increases bleeding in haemophilia. In desperation and fearing for the life of her son, the Czarina Alexandra called upon the more spiritual skills of the peasant preacher Rasputin. He told her to abandon modern treatments and surrender the boy to his own rough brand of faith healing instead. Without aspirin, Alexei's health improved, and Rasputin's power over the Russian royal family grew accordingly. That ascendancy, seen as an increasingly corrupting influence in the febrile atmosphere of the time, had been a significant factor in turning Russians against the Czar and in favour of revolution.

1918, so the realization of ultimate defeat began to sink in back home. When the Allies launched their own all-out attack that autumn and German armies began to buckle under the strain, strikes broke out on the streets of Berlin, mutiny spread through the imperial navy, and the Kaiser was forced to abdicate. With the country imploding, what hope was there for a German businessman, no matter how powerful, who wished to influence events overseas?

On 11 November, Germany surrendered. One month later, as New Zealand troops marched into Leverkusen and confined Duisberg and his family to the basement of their house,* America's APC A. Mitchell Palmer put all Bayer's US interests up for sale to the highest bidder. The indignity could not have been more profound. They were bought for a little over $5.3 million by Sterling Products Inc. – a firm of quack medicine makers.

What a bittersweet irony, then, that Bayer's most popular product was about to face its greatest challenge as the world became enveloped in the most lethal pandemic in history.

---

* In the days immediately following the armistice, it is possible that Duisberg may have skipped town to avoid the wrath of the advancing Allied armies. As one report in the *New York Times* of 24 December explained: 'Dr Carl Duisberg of Leverkusen is reported to have fled to Switzerland. He was generally looked upon as the link between business and General Ludendorff and was one of the most active Pan Germans.' There is certainly some truth in the latter part of this statement, as he had had frequent contacts with Germany's military leaders throughout the war. On 9 September 1916, for instance, Duisberg met Ludendorff and Hindenburg on the train of the Supreme Command to talk over munitions programmes, and the following week he was proposing to them that Belgian workers be used to meet German industry's manpower needs. As defeat began to seem inevitable in August 1918, Ludendorff had even urged Duisberg to lead a delegation to the Kaiser recommending that he abdicate, but Duisberg refused. However, if he did flee in November 1918, he was very soon back at Leverkusen, working to keep Bayer going.

7

CIVILIZATION COULD DISAPPEAR . . .

AT FIRST SIGHT there is nothing particularly remarkable about the cemetery. Surrounded by a high flint wall, it covers a few acres of hillside on the northern edge of the seaside town of Seaford where neat suburban houses begin to give way to the gently rolling countryside of the South Downs. The graves are, for the most part, well tended and unpretentious, their headstones carved with the modest tributes the English have used for centuries to say farewell to their loved ones.

But here and there, clustered in groups of twenty or thirty, are even simpler markers, small white memorials of a kind familiar to anyone who has ever walked across the fields of the Somme. There isn't much on them. A name, rank and service number, a regimental badge or a national motif – a few are British, a few West Indian, most bear the maple leaf of Canada.

What stands out, though, are the dates on which many of these men died: 15 November 1918, 12 December 1918, 22 February 1919 – days, weeks, even months after the armistice that brought the Great War to an end. Clearly, these soldiers lost their lives in the service of their country, otherwise they would not be here. But not all of them died in the war.

So what happened to them? What brought these young men of Ottawa and Montreal and Saskatchewan to a final resting place in Seaford Cemetery on England's Sussex coast?

Stand by their graves for a while and out of the past come the echoes of a childish rhyme, a skipping song such as their pigtailed younger sisters might once have chanted in a playground far away.

> I had a little bird
> Its name was Enza.
> I opened the window
> And in-flu-enza.

The Great Influenza Pandemic of 1918–19 is one of history's forgotten tragedies, a plague out of the Book of Revelations that swept across the world in two devastating waves. It killed at least five times as many people as had perished in the war, indiscriminately harvesting lives from remote communities and big cities alike. For its most lethal form there was no cure, no vaccine, nothing other than your own immune system to ward off attack. If you caught it, you prayed, wrapped up warm and took the few available therapeutic medicines that might help prepare your body for the onslaught to come. Of these, one stood out – a drug that lowered your temperature, eased the aches in your muscles and joints and gave your natural defences a chance to fight back. Aspirin didn't *cure* a single case of influenza, but it helped millions of people in their battle with the virus and undoubtedly saved many lives as a result. After twenty years of court battles, patent wrangles and savage commercial rivalry, this was the drug's coming of age.

History does not relate what drove Albert Gitchell into the army. Perhaps he was dazzled by the same old promises of glory that recruiting sergeants have always dangled before potential volunteers, or maybe he was drafted as a consequence of America's Selective Service Act. Whichever it was, he probably shared the one hope that most young soldiers have when signing up to fight for their nation in wartime – that fate had him marked out for luck and bravery in action, rather than some horrible future as a maimed and bloodied casualty. Of course, any secret ambitions he may have held for battlefield distinction would have dimmed a little when he found out what the army had in store for him – they wanted him to be a cook. But at least there was some short-term compensation. The food and warmth of the company kitchens were much prized in the freezing windswept environs of Fort Riley transit camp during the Kansas winter of 1917–18. And in any case, as it turned out, Private Gitchell *was* destined to become famous, though perhaps not in ways he quite expected.

Shortly before reveille on the morning of 11 March, he lay shivering in his bunk. One of the habitual dust storms howled across the camp's parade ground and the wind was finding its way into the cracks around the windows. The stove was out and Gitchell was finding it impossible to stay warm. In a moment he had to be up, dressed in fatigues and making his way across to the cookhouse to prepare breakfast, but he ached so much he could barely move. Eventually he managed it and struggled outside, coughing in the cold air. When he got to the kitchens, fires were already being lit and breakfast was under way. He did his best to join in, but after a few fumbled attempts to break eggs, an exasperated NCO sent him over to the stove to warm up with orders to report to the camp hospital as soon as it opened.

When he finally got there an hour later Gitchell was feeling much worse. He was running a fever, his throat ached and he had a raging headache. The duty sergeant medic, a stern old veteran, was on the lookout for malingerers, but one brief examination was enough. He gave Gitchell two Bayer Aspirin tablets and ordered him to bed in a ward reserved for contagious diseases. Barely had the cook left when another man appeared, a Corporal Lee W. Drake of the First Battalion's Headquarters Transportation Detachment. He had identical symptoms. Then came Sergeant Adolph Hurby with a temperature of 104 degrees. Another man followed, then another. The medic began to get worried and called in some help. By the time the camp doctor arrived, the line of sick men stretched outside the infirmary into the parade ground. At the end of the day 100 soldiers had been confined to their beds. By the end of the week that number had jumped to 500, all suffering from the same complaint.

Thus, by a few minutes, Private Gitchell became the first recorded victim of the Great Influenza Pandemic of 1918–19. He survived, but by the time the pandemic had fully run its course more than 50 million others would have died.

How the virus had found its way to a US Army camp is open to conjecture. Although it was clearly very infectious, the early cases weren't treated all that seriously. Flu was just one of those things that people caught every winter and this particular strain didn't seem to last too long. Even the forty-eight fatalities at Riley that spring didn't give huge cause for concern; they were spread over several weeks and it wasn't an unusually high mortality rate, given the thousands of soldiers who contracted it. Most of the victims were confined to their beds for three or four days and then were back on their feet. The US Army recorded the outbreak, because keeping records is one of the things that armies do, but at the time no one thought to ask too many questions about where the virus might have come from. Later, of course, scientists would frantically try to piece the puzzle together (some are still trying today), and various theories emerged. Some said that it originated in China and that an earlier, milder version had undergone a genetic mutation while being carried by immigrants to the United States. Others have since come to believe that it was a variant of swine fever, a type of influenza that had lain dormant in pigs for many years until it developed into a killer strain that leapt across species.

What mattered most in those first weeks, though, was its infectiousness, and the United States in early 1918 was an ideal place for it to spread. The country was gearing up to fight. Its small peacetime army had expanded dramatically in the months following the declaration of war on Germany, and tens of thousands of new recruits and draftees were crammed into vast camps for basic training before being shipped overseas. There were 26,000 in Fort Riley alone. Once the flu got a hold in one of those camps, it was only a matter of time before it leapt into others. Forts Hancock, Lewis, Fremont and Sherman all began to report cases and the virus even penetrated the infamous San Quentin prison in California where 500 inmates came down with the same condition. However, for the time being, it was still largely an institutional problem, a three-day contagion spreading among young men living in close quarters, and public health officials noticed nothing untoward in the civilian population.

But March 1918 was when the mass movement of American troops to Europe began in earnest. That month 84,000 soldiers left. In April, 118,000 followed. And when they sailed, they took the virus with them. The 15th US Cavalry suffered thirty-six cases in its trip across the Atlantic. Six men died, an alarmingly high number in the early stages of the epidemic, and although the primitive nursing facilities on board ship were probably as much to blame, it was an ill omen for the future.

Thus influenza arrived on a continent battered and wearied by almost four years of war, and it was as though someone had lit a fuse. At first it played havoc among the American troops fanning out from the Channel ports to their allotted positions at the Front. Then it spread to the French and British trenches and leapt across no man's land to the German lines. Tens of thousands of men began reporting sick on both sides. Germany's General Erich von Ludendorff, trying to maintain the momentum of his massive spring offensive, complained that the flu was having a devastating effect on the morale of his soldiers. 'It was a grievous business having to listen every morning to the Chiefs of Staff's recital of the number of influenza cases, and their complaints about the weakness of their troops,' he would say later. But his opponents were hit just as hard. The British Army's 29th Division had to postpone a counter-attack because its ranks were so devastated by the flu. The 2nd Battalion of the King's Royal Rifle Company recorded in its war diary how it had to establish a temporary hospital in abandoned school buildings at Noeux.

They were not alone. Virtually every unit and regiment in every combatant army that late spring had its own mini epidemic, the massed ranks of cold, wet and often exhausted men absorbing and incubating the virus before passing it on. And inevitably, it came out of the trenches and into the civilian population; racing south through France to neutral Spain where one third of the population of Madrid fell ill, government offices had to close, and newspapers, unhindered by wartime censorship, filled their pages with reports of its virulence. The Allied press, precluded by censorship from reporting the spread of the flu in their own countries, repeated these stories, which later led to the whole epidemic being dubbed the Spanish Influenza. This was more than a little unfair because at the same time, the virus was also moving west across the Channel to Britain (where George V caught it and the Grand Fleet could not put to sea for three days because 10,000 sailors were ill), and east through Germany to Poland and the thawing muddy wastes of the old battlefields of the Russian Front. In fact, much of the world caught the flu that spring, from Europe to the Middle East to Asia, although for now at least Africa, South America and, strangely, much of Canada remained flu-free zones.

And then, as suddenly as it had arrived, it seemed to vanish. The fevers came down, the aches dissipated, the sore throats returned to normal and everyone breathed a sigh of relief.

But the virus hadn't gone. It was just gathering strength, mutating amid the trenches and in the rest billets and the hospitals of the Western Front, growing in the very lungs of the armies that were preparing for the final climactic battles of the Great War and quietly making its way on to the convoys and liberty ships of wounded soldiers returning across the Atlantic, ready to infect fresh victims in its newer, more deadly guise.* The second wave, the killing wave, was about to begin. For a few dozen Canadian soldiers, the graves in a quiet English cemetery had suddenly come a step nearer.

The US Navy sailors who docked at Boston's Commonwealth Pier in August 1918 were in transit, looking forward to a few days' shore leave before being reassigned to other ships coming fresh off the slipways in the city's dockyards. As they spread out among the bars along the waterfront, glad to be away from the dangerous monotony of convoy duty, few would have given much thought to the flu. But then, one by one, they began to fall ill. On 28 August, ten of them got the flu. The following day another fifty-eight reported sick and were transferred to the Chelsea Naval Hospital. By 7 September, just over a week later, 119 of them had gone down with the disease and Boston City Hospital reported its first civilian case. On 8 September, they started dying. The influenza was back with a vengeance.

To Hungarians it was the Black Whip, to German troops it was known as *Blitzkatarrh*, the Swiss dubbed it *La Coquette*, and the rest of the world picked up on newspaper references to the Spanish Influenza or stuck with *La Grippe*, the name traditionally given to the flu in Europe. But whatever you called it, this new strain of the disease was like nothing anyone had ever seen before. If you were lucky, you were one of the 20 per cent of victims who caught a milder three-day fever, similar to that of the first wave. It was unpleasant, you felt dreadful, but you got over it.

For many of the rest, though, it would become a life-and-death battle. It might start as ordinary flu, but instead of recovering, your aches and pains and chills would get worse, your breathing would start to labour and your shivering would become uncontrollable. By day five, your damaged lungs were being ravaged by swarms of deadly bacteria and pneumonia began to overwhelm you. And then it became a race between the disease and your body's capacity to fight back. Even if you survived, it would take weeks of convalescence before you were fully functional again.

---

* One further extraordinary theory about the second wave, put forward in 1948 by a New Zealand scientist, Dr G. M. Richardson, was that the widespread use of mustard gas by both sides in May and June 1918 might have released chemical compounds that increased the virus's toxic mutation.

Others didn't even have that slim chance. Hours after they caught the flu, their lungs would begin filling with fluid, cutting off their oxygen supply. They would start coughing up blood and sputum, their faces turning blue as they gasped frantically for air. Delirium would set in and then unconsciousness. And then, quite literally, they would suffocate and drown. From catching the flu to being killed by it could take as little as twenty-four hours.

That was the experience of many soldiers at Fort Devens, Massachusetts, thirty miles from Boston. The same day that people began dying in the city, the epidemic took hold in a US Army camp crammed with 50,000 young draftees. It was as though the grim reaper strode among them with his scythe.

Sixty years later, a letter written by one of the Fort Devens doctors was found in a tin trunk by a Dr N. R. Grist from the University of Glasgow. He sent it to the *British Medical Journal* for publication. The letter is dated 29 September 1918, three weeks into the outbreak, and bears the simple signature 'Roy'. After describing to his correspondent how normal life in the camp had ground to a halt, Roy explained what had been happening to the soldiers reporting to the hospital with flu.

> Two hours after admission they have the Mahogany spots over the cheek bones and a few hours later you can begin to see the Cyanosis extending from the ears and spreading all over the face, until it is hard to distinguish the coloured man from the white. It is only a matter of hours then until death comes and it is a simple struggle for air until they suffocate. It is horrible. One can stand to see one, two, or twenty men die, but to see these poor devils dropping like flies gets on your nerves. We have been averaging about 100 deaths a day, and still keeping it up . . .

Alarmed by reports from Fort Devens, the government sent in an eminent team of doctors to investigate. They arrived in the pouring rain, a few days before Roy wrote his letter, and found a scene straight from a medieval depiction of hell. Thousands of soldiers sat coughing and shivering in their tents, whilst long lines of others, festooned in greatcoats and damp blankets, queued outside the hospital gates. Inside the building, designed to hold 2,000 patients, 8,000 men lay in the grip of pneumonia. Sixty-three of them were to die that very day. One of the doctors, Colonel Victor Vaughan, a past President of the American

Medical Association, and the Army's Surgeon General, wrote later that the memories were ghastly ones 'which I would tear down and destroy were I able . . .' Another of the team, Dr Rufus Cole of the Rockefeller Institute, was startled to discover that he had to climb over the stiff bodies of the dead to get into the autopsy room.

Appalled by what they had found, the investigators telegraphed the US Public Health Department with a frantic plea for more doctors and nurses. But little help would be forthcoming. By then the disease was everywhere. Already 50,000 people in the state of Massachusetts had the flu and it was spreading like a bush fire around the country, from army camp to army camp, from city to city. At the end of September, 12,000 Americans had died and hundreds of thousands more were seriously ill. And that was only the beginning. As the sickness tightened its grip on the United States, so it began to reach out across the world with all the speed and power of a giant tsunami. Within two weeks it had appeared in every part of the globe. From then on, in every town, every village, almost anywhere where people met, from Chicago to Cape Town, from Karachi to Canterbury, the deadly contagion would strike down its victims.

On 30 September, the *Newfoundland Evening Telegram* reported that three seamen had been admitted to hospital in St John's. Two more were admitted the following day. The deaths began the following week. In Glasgow, the flu began to roar through tenement blocks and back alleys with a ferocity not seen since the cholera epidemic of 1849. In Lagos, Nigeria, ten sailors with flu infected the whole city within a fortnight. In Buenos Aires the overcrowded hospitals began barring their doors to new arrivals. Hundreds of Allied troops, disembarking at Archangel in support of White Russian forces fighting the Bolsheviks, stepped straight off the boats into their sickbeds; a mere six days after landing the US First Battalion had suffered twenty-four fatalities and in Archangel itself, 10,000 went down with the flu. On and on it went, sweeping across continents, leaping across seas to Berlin, Shanghai, Tokyo, Cape Town, Oslo. The rapidity with which the disease spread was astonishing – it was estimated by Rhodesian doctors to be travelling at thirty-five miles an hour when it hit their country at the beginning of October. One week it was rushing through six adjacent workers' cottages in the tiny Irish village of Letterkenny, the next it had somehow found its way to the Maatsuyker Lighthouse, six miles off the coast of Tasmania, where the keeper and his wife hadn't seen a soul for three months.

In part, this incredible reach was a consequence of the mass movement of men in wartime, in addition to an already sizeable stream of merchant shipping around the world. It was also due to ignorance about the importance of quarantine during an epidemic, knowledge that was often only acquired when it was too late.

The story of how it hit Fiji and New Zealand is a perfect example of both these factors. At 11.50 a.m. on 11 October 1918, Captain John Rolls, of the Union Steamship Company's vessel *Niagara*, sent a telegram to the Admiralty office in Wellington.

> Please advise Health Department Spanish influenza cases aboard; increasing daily. Present time over 100 crew down. Urgently require hospital assistance and accommodation for 25 serious cases.

The 13,000-ton ship was in passage from Vancouver in Canada with 300 passengers and crew. The first signs of the virus had appeared just three days out when a cabin boy fell sick. By 8 October there were eighty-three cases on board, and the ship's doctor, who had at first mistaken the mini epidemic as an outbreak of dengue fever, was feeling the strain. The following day the *Niagara* put in briefly to Suva on the island of Fiji (where, incredibly, a few of the crew were allowed on shore) and then set off on the final leg to New Zealand. The first on-board fatality came soon afterwards.

It would only have been common sense for the *Niagara* to have flown a quarantine flag as she approached New Zealand. The ship was a hotbed of infection that should have been contained. But she was due to dock in Auckland harbour the following day and Captain Rolls had VIPs breathing down his neck. Among the passengers were New Zealand's Prime Minister, William Ferguson Massey, and the country's finance minister, Sir Joseph Ward, desperately anxious to get back to their desks after a five-month absence at an Imperial War Conference. Whether this influenced the country's Health Department isn't clear, but to the fury of the local dockers' union, which had picked up the news and was threatening to boycott the ship if she came in, the *Niagara* was allowed to berth. Police constables were on hand to carry the very sick to ambulances and police cars, while Massey and Ward climbed into an open landau and proceeded to parade, top hats a-waving, through the Auckland crowds that had come out to welcome them home. So influenza arrived in New Zealand and eight weeks later 8,251 people had died.* The consequences of the *Niagara*'s brief stop over in Suva were almost as bad. Eight thousand Fijians would die of the flu.

And everywhere it went, the outcome was the same: devastation, death and a medical community punch-drunk from the appalling savagery of a disease they couldn't understand.

In Britain, *The Times* reported:

> Yesterday there were 1,445 members of the Metropolitan Police and 130 members of the London Fire Brigade on the sick list with influenza. During the 24 hours ending at seven o'clock yesterday morning, forty-four persons were stricken with sudden illness in the London streets and were removed to hospital in the LCC ambulances. In Battersea undertakers have been compelled to refuse to take orders for funerals. One undertaker has declined twenty orders.

The pandemic was becoming a plague of biblical proportions, a human catastrophe of a size and scale not seen since the Black Death. As the death toll mounted, stunned public health officials were besieged by questions for which there were no apparent answers. Where had it come from? What could be done to stop it? How bad would it get?

It is ironic, given the role that it played in helping people cope with the flu, that aspirin was actually blamed by some for starting it. Shortly after the virus's appearance in Boston, the rumour went around that influenza germs had been deliberately inserted into Bayer Aspirin as a new and dreadful kind of chemical weapon. If you took one to stave off the aches and fevers, you became infected instead. The fact that the epidemic was just then beginning to affect Germany in exactly the same way as the United States, didn't seem to have occurred to anyone at the time. The war was just entering its last phase, the newspapers were full of a new American attack on the Siegfried Line north of

---

* These were not the first New Zealanders to die. Some of the 40th Reinforcement of the New Zealand Army had caught the flu on board the troopship *Tahiti* when she stopped at Freetown, Sierra Leone, en route to Plymouth some weeks earlier.

Verdun, and anti-German rhetoric was the order of the day. Another rumour had it that at Camp Hancock, New Jersey, a group of army doctors and nurses had been executed by firing squad, as German spies, for spreading the flu by hypodermic shots. These were insane times, when even supposedly responsible government officials contributed to the hysteria. On 21 September Lieutenant Colonel Phillip Doane, head of the Health and Sanitation Section of the Emergency Fleet Corporation, told the *Philadelphia Inquirer* that,

> It is quite possible that the epidemic was started by Huns sent ashore by Boche submarine commanders. We know men have been ashore from such submarines. It would be quite easy for one of these German agents to turn loose influenza germs in a theatre or some other place where large numbers of persons are affected. The Germans have started epidemics in Europe, and there is no reason why they should be particularly gentle with America.

US Bayer's new board quietly stepped up its advertising which emphasized that its aspirin was now all American made, but it took a long while for the rumours to stop circulating.

The sad truth was that public health officials were at a loss as to how to respond to this disaster. The city of Philadelphia was a case in point. With a vast naval yard and several army transit camps within fifty miles, the city should have been on full alert for the possibility that infection would set in, but its small public health office was completely unprepared for what was to come. The first cases appeared on 11 September, yet ten days later 200,000 people were encouraged to crowd into the city centre to see a huge Liberty Loan Drive parade. Apart from a campaign to warn against the dangers of spitting, coughing and sneezing in public, nothing had been done to prepare them. This kind of complacency had a price. Barely a month later, 11,000 people had died and tens of thousands more were sick. By then hospitals, doctors and undertakers were overwhelmed, rotting bodies were piled up three deep in the city morgue, and bereaved families were soon having to dig graves for their loved ones themselves. Coffins were in such short supply that some corpses were just abandoned in the street. People began to avoid houses that were suspected of containing a flu victim. It could not have been more reminiscent of the medieval bubonic plagues had city officials gone around daubing crosses on the front doors of infected homes, shouting 'Bring out your dead.'

Although Philadelphia was one of the worst-hit US cities, similar scenes were being played out across America. In New York, the health commissioner, Royal Copeland, announced at the end of September that, 'the city is in no danger of an epidemic. There is no need for our people to worry.' A couple of weeks later 851 New Yorkers died of influenza in a single day (contributing to a death total in excess of 33,000) and steam shovels were being used to dig temporary mass graves. The equally bullish chief of San Francisco's Board of Public Health predicted that flu wouldn't even *reach* his city. One month later over 2,000 of its inhabitants were dead.

Once flu set in, of course, the authorities responded as best they could. The American Public Health Association issued a stream of recommendations mostly aimed at prohibiting unnecessary assemblies. Public meetings and sports events in many parts of the country were banned, schools were closed, barracks quarantined, stock markets, banks and shops shut, trains, trams and buses brought to a halt. Even churches were asked to keep services to a minimum. Huge drafts of soldiers, much needed on the Western Front, were postponed. Gauze masks were made mandatory in some cities. In Rockford, Illinois, they were rather dramatically emblazoned with a black skull and crossbones. San Franciso's ordinance was accompanied by the ditty: 'Obey the laws and wear the gauze / Protect your jaws from the septic paws.'

Similar regulations were issued in other countries. In Britain (where in London alone, 14,000 people would die in the first week of November), the same strictures on public assembly were issued, although they didn't always meet with universal approval. As one editorial in the *British Medical Journal* pointed out, 'every town dweller who is susceptible must sooner or later contract influenza whatever the public health authorities may do, and the more schools and public meetings are banned and the general life of the community dislocated the greater will be the unemployment and depression'. However, almost everyone began to wear face masks when they went out and it became commonplace to see council workers, armed with hand-held sprayers, parading through the streets squirting fellow pedestrians with clouds of disinfectant.

But this was not a gentle bug that could be treated with a few drops of carbolic, or even, as was becoming obvious, by complete isolation.

A virus that could find its way to the Arctic wastes of Alaska and wipe out whole Inuit villages, or could kill 20 per cent of Western Samoans in a matter of weeks was clearly capable of extraordinary feats of transmission. Nor would it be enough, as Britain's *News of the World* suggested in all seriousness on 3 November, to 'wash inside your nose with soap and water each night and morning; force yourself to sneeze night and morning, then breathe deeply; do not wear a muffler; take sharp walks regularly and walk home from work; eat plenty of porridge.' No, what everyone expected was that medical science would be able to come up with a medical response. Vaccines had been found for smallpox and anthrax, so why wasn't one being developed for influenza?

The simple explanation was that the scientists (many hundreds of whom had indeed begun searching desperately for a solution) didn't appreciate what they were up against. Just as eighteenth-century English doctors had believed that stagnant waters caused the ague, whilst ignoring the biting swarms of malarial *Anopheles* mosquitoes that flew above them, so their modern counterparts failed to see what was in front of their eyes. They knew that the disease spread through the air, but the microscopes of the time were not powerful enough to detect the tiny virus that caused the flu (and would not be until 1933).* In fact the very notion of a virus as an infectious parasite had not yet been coined. But early twentieth-century medical science did understand bacteria and because it had previously been able to produce bacteriological vaccines for such things as typhoid fever, smallpox and tetanus, the natural assumption was that an influenza bacillus existed and could be used to provide preventive serums. The only problem was that none of them worked. A Dr C. Y. White of Philadelphia announced that he had developed one in October 1918, and 10,000 sets of inoculations were prepared for the Board of Health, but it made no apparent difference to the rate of infection throughout the city. A similar attempt to vaccinate British troops in France was equally unsuccessful.

And in the meantime, the mortality rate was climbing around the world. In the United States alone, October saw a staggering 195,000 deaths. Colonel Victor Vaughan, the Army's chief doctor, reached a frightening conclusion: 'If the epidemic continues its mathematical rate

* Around a billion influenza virion could fit on a match-head.

of acceleration, civilization could disappear from the face of the earth in a matter of weeks.'

With no effective vaccine forthcoming, doctors had no choice but to fall back on old-fashioned, common-sense medicine – good nursing, isolation, hygiene, and simpler therapeutic remedies that might at least boost some people's chances of fighting off the infection. In October, the US Surgeon General, Rupert Blue, briefed the press on how to recognize influenza symptoms early and then gave America's ailing victims the only prescription he could – stay in bed, eat as well as you can to keep your strength up, and take aspirin.

Aspirin's remarkable ability to reduce high temperatures had already become one of its most famous attributes. No one knew exactly how or why it did this (and they still don't know exactly why today) but it had intrigued scientists ever since 1758 when the Reverend Edward Stone had observed this effect in willow bark, the drug's natural ancestor. It wasn't totally unique as aspirin shared this property with quinine, but quinine did not have aspirin's analgesic and anti-inflammatory powers. For flu victims, whose early symptoms included dangerously high temperatures as well as brain-splitting headaches and aching limbs, it was this combination of qualities that made aspirin of such tremendous therapeutic value. Whilst it could do nothing to prevent the onset of influenza, or directly attack any potentially deadly secondary infections like pneumonia or bronchitis, it could at least allow the body a breathing space in which its own natural defences might rally and begin to fight back – providing, of course, that it was given sufficient time; millions of people who died in the 1918–19 pandemic became so ill, so quickly, from such a wide-ranging and dramatic assault on their bodies, that no amount of aspirin could have done any good. However, for many people who clung on longer, their fate hanging in the balance, the drug could give them that small chance that mattered, even if the most it did was to make them feel a little better. There is no way of proving scientifically that someone's will to live can help decide their fate, but few doctors will deny that once everything else has been tried, the strength of a patient's instinctive determination to survive can make the difference between life and death. By making very sick people a touch more comfortable, aspirin helped them gather that strength.

Not that everyone recognized this at first, of course. When no official 'cure' was forthcoming, people turned back to old folk remedies and even invented a few new ones – garlic essence, mothballs, kerosene on sugar, powdered cinnamon and eucalyptus oil were among the most popular. Smoking, believed by some to ward off infection, was actively encouraged, even in British munitions factories, and was actually made compulsory for employees of a chain store in Zwolle, Holland.* Many of the more scientifically minded took to gargling with antiseptic on the basis that this might disinfect their airways; others took refuge in the products of the patent medicine business – in just one month 56,612 bottles of a popular American cold remedy called Taniac were sold. The cranks and charlatans had a field day too: gullible Louisianans bought 'sacred pebbles' that had supposedly been blessed in a shrine in Japan. One North Carolinian recommended sprinkling sulphur in your shoes and strapping slices of cucumber to your ankles. More credibly, quinine was in great demand, despite being in short supply and not particularly effective. Bayer's other old medicine, Phenacetin, which had made its reputation during an earlier, milder epidemic at the end of the nineteenth century, also sold very well, although a Philadelphian doctor, J. M. Anders, campaigned vigorously against its use on the grounds that it could cause nephritis.

So, increasingly, the medical community turned back to aspirin, which seemed the most effective (or the least damaging) of all the therapeutic treatments available. Inevitably, though, there was confusion about how it should be dispensed. In Delhi, the eminent Indian surgeon Dr Manjunda Rao insisted that smart young doctors in Bombay were recklessly misusing it – on the basis, he said, that it weakened the heart (an old canard) and actually brought on pneumonia as a consequence. In London, a Harley Street specialist, Dr Edward Turner, disagreed: his prescription called for a patient to be 'drenched' with aspirin: twenty grains an hour for twelve hours non-stop, and every two hours thereafter. (At doses that high, of course, the patient was likely to suffer severe intestinal bleeding.) A fellow countryman was equally effusive. Dr Robert Jones, from Lancashire, said that it was possible that aspirin might actually be the elusive cure that everyone was looking for – or at least that there was something in the salicylates

---

* Interestingly, the one person who ignored this instruction was the only one who didn't get the flu.

that prevented the spread of the disease. There wasn't, but it was a sign of the drug's growing status. When influenza reached Australia (it would claim 6,387 lives in New South Wales alone) the government took the usual precautions – making masks compulsory in public places, closing down places of public assembly and enforcing quarantine rules on the infected – but also made use of a wartime regulation giving it the power to fix the price of 'necessary commodities'. Aspro, the Nicholas brothers' new all-Australian aspirin, was at the top of the list.* In New Zealand, one Maori village was so impressed by the power of the drug that it bestowed a singular honour on F. G. Wayne, the local health superintendent who had been giving it to them. On one of his visits they invited him to attend a christening party, where, to his astonishment, he found that the baby had been named Aspirin Wayne. Sales were rocketing everywhere. In Camp Sherman, Ohio, Major Carey McCord included an order for 100,000 aspirin tablets among his urgently needed medical supplies. In Paris, the demand for aspirin was so acute that it was despatched to pharmacists under police supervision.

But the numbers of sick and dying were increasing everywhere too. The disease was no respecter of class or rank. Those who fell very sick but survived included Assistant Navy Secretary Franklin D. Roosevelt, who collapsed on board the battleship *Leviathan*, and the film star Mary Pickford, the richest woman in the world, who was struck down in Beverly Hills. Queen Alexandrine of Denmark caught it, as did Brazilian President Wenceslas Braz and the Sultan-Caliph Mohammed VI of Turkey. The British Prime Minister, David Lloyd George, spent a feverish week chewing aspirin in a Manchester hotel room; Crown Prince Max of Baden, the last Chancellor of imperial Germany, went down with it a few days before the Kaiser's abdication. The Kaiser himself, who at lunch with his generals on 1 October had raised the hope that the flu would cripple the Allied armies whilst leaving his own unaffected, also succumbed. US President Woodrow Wilson had caught it earlier, so badly that some historians would later credit his inability to ensure a just settlement for Germany at Versailles to the weakening effects of his long convalescence.

Others were not so lucky. Sir Hubert Parry, composer of 'Jerusalem'

* Much to the delight of the Nicholases' new advertising guru, George Davies.

died at Rushington in Sussex. The Dowager Queen of the Tongan Islands died. Silent screen star Harold Lockwood died. General Louis Botha, the first Premier of the Union of South Africa, died. The Maharajah of Jodhpur; Edmond Rostand, author of *Cyrano de Bergerac*; and Admiral Dot, one of Phineas T. Barnum's midgets – all of them died.

And for every supposedly noteworthy victim there were tens of millions of more ordinary people who succumbed. The death toll in India alone was staggering. Estimates of the mortalities there have varied because of the inadequacy of the record-keeping, but even conservative assessments now put the number at between 16 and 18 million people. That is twice as many people as died in combat during the grotesque slaughter of the First World War. The statistics for Africa – aside from South Africa where around 140,000 died – are largely unknown but were possibly in the tens of millions. Whilst it is perhaps understandable why the figures were so high in non-industrialized countries, which had only the most primitive medical care, and totally inadequate supplies of even basic therapeutic medicines like aspirin, they are no less shocking for that. India's plight was worsened considerably by the fact that many of her limited number of trained doctors had been sent overseas with the Indian Army. In parts of tropical Africa and South East Asia (where an estimated 1.5 million died in Indonesia alone) flu was an almost unheard-of rarity and there was little natural immunity for victims to call on. The same was true amongst the Inuit population in Canada, where 80 per cent of those who contracted the disease died.

In the industrialized world the mortality levels were lower, but still startling. In the United States an estimated 550,000 died (43,000 of them in the services) and the average life span fell by twelve years for 1918. In Britain 228,000 died, and in Germany, 400,000. France had around 300,000 fatalities (excluding flu cases among the armed forces of both sides). In Italy, where the proto dictator Benito Mussolini was so shocked by the fatalities that he called in his newspaper *Il Popolo d'Italia* for handshakes to be made illegal, around 350,000 died. And so the death toll stacked up. After a while, the statistics become almost meaningless, the sheer scale of the pandemic too dreadful to take in. Some historians actually put the total mortality figures worldwide at around 100 million on the basis that death rates for Africa, China, much of South East Asia and South America were never accurately collated and so were almost certainly underestimated. But even if it was only half that number, it is still an astonishing figure. The important thing to remember is that all these deaths were *in addition* to the numbers who would have died from old age, cholera, cancer, starvation or any of the other causes of human mortality.

In the midst of this catastrophe, the end of the war came and went, celebrated most surreally on the streets of San Francisco by a 30,000-strong crowd, all still dutifully wearing their face masks. The rest of the world did their best on Armistice Day to forget that out there an unseen killer still lay in wait, beginning a strange process of mental suppression that would eventually see memories of the worst epidemic in human history expunged from the consciousness of later generations.

How that happened has deeply puzzled the modern historians who have re-examined this astonishing event. Alfred Crosby, one of the first to make a serious study, argued that because the disease was so enmeshed in people's minds with the bloody horrors of the war that preceded it, no one wanted to think about it once that war was over. It is also probably the case that the sheer global scale of what was going on was not generally apparent to the public at the time. Wartime censorship restricted the reporting of civilian illness and death in many countries because this was felt to be strategically important information. In some countries like Italy, even funeral marches were officially frowned on lest the sheer number of them gave the enemy any idea that the country was in crisis. When the war was over, these habits and restraints took a little while to shake off.

The mass movements of people and the social upheaval that the war entailed also distorted the picture. One and a half million American troops went overseas in its last year and millions more Germans, Austrians, French, Australians, British, Italians, Indians, Belgians, Hungarians, Bulgarians and Turks had been displaced hundreds or thousands of miles away from home as combatants, essential workers or refugees. This massive social dislocation did not make for a coherent collective experience and thus the war and the pandemic became part of the same confused jumble.

The young Canadian soldiers who lie in Seaford Cemetery on England's Sussex coast were part of this mass movement. For much of the war, the town was home to two massive transit camps through which tens of thousands of Canada's troops passed to and from the Western Front. The flu hit those camps several times during 1918 and even though the death toll wasn't remarkably high in comparison with elsewhere, it continued to claim victims up to and beyond the end of hostilities. Some of those who are interned there now never even made it to France; men like Sapper McCallum of Rodney, Ontario, who was mobilized for service with the 3rd Canadian Engineer Reserve Battalion but who died from the flu in Seaford instead. Others succumbed on their return from the war, while waiting to go home for demobilization. What a tragic irony that must have been – to have survived the horrors of Ypres, the Somme and Verdun, to be in sight of peace and a hero's welcome, and then to die from the flu in a small English seaside town. Of course, many of them would have suffered the same fate at home; more than 50,000 Canadians lost their lives in the later stages of the epidemic. No wonder that people saw the war and the flu as part of the same ghastly nightmare and did their best to forget both.

These were among the last few tens of thousands of Western fatalities. The pandemic relit briefly on the other side of the world over the summer of 1919, when Australia relaxed its quarantine rules and the disease flared up again, but it was running out of vulnerable victims. This particular form of death had lost its sting. When it was over, the medical community could make some kind of accounting. It had made some fatal, if understandable, errors and had learnt some lessons as well; basic things like the importance of quarantine, hygiene and good nursing; giving the simple principles of public health the attention they deserved. As for the killer virus, it vanished over the winter of 1919–20 never to be seen again, although scientists have worried ever since that it might one day reappear.*

In the meantime, the role that aspirin had played would linger among the half-suppressed memories of many of the survivors of the 1918–19 pandemic. The tragedy had cemented the drug's reputation in a way that almost nothing else could have. Of course, it hadn't stopped the disease, but people knew that it had been tremendously helpful, one of the few medicines to make any difference, and in the darkest moments that had been something to cling to. Production and sales of aspirin

* The global SARS outbreak of 2003 briefly rekindled those fears, although fortunately it turned out to be much less infectious.

more than doubled between 1918 and 1920 as manufacturers went into overdrive to meet the huge demand. Many millions of people who had never tried it before had become convinced of its therapeutic powers, and having used it once they would want to use it again and again.

This potential of this massive new market was not lost on those who made aspirin, nor did they forget that a world at peace was a world with money to spend. Normal service was about to be resumed. Even as the nightmare of the epidemic and the war was ebbing away, rival producers were marshalling their own forces. The aspirin age was about to begin.

8

THE ASPIRIN AGE

GERMANS HAD FEW reasons to celebrate in the winter of 1918, although people stoically tried to make the best of a bad situation. Some greeted the armistice with much the same relief as the rest of the world, finding it impossible to remain unmoved by the news that four years of appalling slaughter had come to an end. Others took comfort from the fact that on 11 November the German Army occupied roughly the same positions as it had in August 1914. Germany had been defeated but not totally conquered, her borders remained largely intact and her battered economic infrastructure still just about functioned. Things could have been worse, people told one another. Things could have been much worse.

But over the following weeks and months the cold, depressing, inescapable reality of defeat began to sink in. Over 1.7 million German soldiers had been killed or wounded for no apparent gain; food, already heavily rationed, was in increasingly short supply; and the flu pandemic still raged in German cities and among the angry and demoralized ranks of her slowly returning army. Moreover, the country's political structures were on the verge of collapse. The new socialist republican government, in uneasy alliance with the military hierarchy, had forced the abdication of the Kaiser and had successfully negotiated the surrender, but the Chancellor, Friedrich Ebert, still had only the most tenuous hold on power. A Spartacist revolt in Berlin that winter was suppressed only with the help of the Freikorps, a hastily assembled force of former servicemen, and elsewhere soldier soviets and anarchists continued to threaten revolution. In the meantime, there was growing uncertainty over what conditions the Allied powers, whose troops now occupied the west bank of the Rhine, would seek to attach to any final peace settlement.

For Carl Duisberg, temporarily living with his family in two rooms in the basement of his grand house, things looked bleak too. He wasn't one of nature's pessimists yet his habitual self-confidence must have been sorely tested that winter. With his colleagues on the Farbenfabriken Bayer board he tried to engender a calm transition back to peacetime production, and to endure as best he could the unwelcome presence of the New Zealand troops garrisoned at the Leverkusen plant, but the political chaos outside the factory gates made things difficult. By the start of 1919 output levels had fallen to about 60 per cent of what they had been in 1914.* Gradually, as the foreign soldiers left and the political situation improved, things would get better, but it would take all Duisberg's determination to re-establish Farbenfabriken Bayer and the broader IG Farben combine as the force they had once been. In the meantime, one of his biggest problems was that many of the company's most precious assets had been confiscated and it would be hard, if not impossible, to get them back. Included among these, of course, was aspirin, the jewel in Bayer's crown, which even then was proving its worth as the key palliative treatment for influenza. Germany's former opponents had learnt how to make the drug for themselves and there was nothing to stop them profiting from that knowledge or from the increased demand for the drug. That rankled hugely with Duisberg.

The situation in the United States was particularly galling. All of Bayer's US assets had been sold off in a shabby public auction to a company that appeared to be the very antithesis of everything that Duisberg stood for. He was never afraid to make money, but he had always insisted on the finest science and the highest quality standards for Bayer pharmaceuticals. Now the most important of those drugs was to be produced in America, under Bayer's name, by a company that was best known for its tawdry patent medicines.

Sterling Products Inc., of Wheeling, West Virginia, was the brainchild of two small-town pharmacists, William E. Weiss and Arthur Diebold. Childhood friends from Canton, Ohio, they came together in 1901 to cash in on the booming patent medicine trade and set up an outfit called the Neuralgyline Company. Its only product was a quack analgesic known as Neuralgine, which, in keeping with the down-home traditions of their chosen profession, they sold from the back of a

* With the exception of aspirin. Its sales doubled.

horse-drawn buggy. But despite these humble beginnings, the two quickly proved to be highly effective salesmen, racking up a $10,000 profit in their first year of business, most of which they pumped straight into newspaper advertisements for their tonic. It was a clever and successful strategy; more Neuralgine was sold, more profits were made, and more advertising space was purchased. As the company grew, so the partners began to buy up other patent medicine firms, including Sterling Remedy, which produced a remedy for nicotine addiction. To these they applied the same sales–advertising–sales formula as they had with Neuralgine and again it worked. By 1912 the combined business was worth over $4 million, had changed its name to the more respectable sounding Sterling Products Incorporated and was looking around for other ways to expand.

When the Office of the Alien Property Custodian announced it was to sell the Bayer company in November 1918, the two men saw their chance to lift their ambitious but medium-sized firm into the big league. They entered the auction and, a little to their surprise, secured the American assets of one of the world's most important chemical companies for a little over $5.3 million. This was a large sum for the day and they only just managed to beat off other better-known names like Du Pont and Paine Webber, but in return Sterling got Bayer's massive Rensselaer plant, its dyestuffs business and US sales and production rights to sixty-four of its most scientifically impressive medicines. Most importantly of all, the American trademarks came with them. Among them was Bayer Aspirin, one of the hottest pharmaceutical products on the market.

Weiss and Diebold had no interest in the dye business, which they quickly sold off. But they had big plans for the rest. Sixty-three of the pharmaceuticals, including brand names like Salvarsan and Phenacetin, were bundled together into a newly formed subsidiary called the Winthrop Chemical Company – a move designed to obscure the German origins of the drugs at a time of still considerable ill-feeling. Aspirin was kept out of the arrangement, though. Bayer's name was now so powerfully connected to the drug that changing it could damage the product's high profile, and in any case Weiss and Diebold believed that any lingering Teutonic connotations it might have would soon be washed away in the big marketing campaign they had planned. Sterling Products had made its fortune through advertising its quack remedies and now it would apply the same strategy to one of the world's most successful medicines. Weiss, who as president managed the company on a day-to-day basis (Diebold looked after the money), told his board that up until then 'the field has been merely scratched on the surface and there are tremendous possibilities ahead'.

The implication was clear. Aspirin was going to be sold with all the hype, fanfare and panache that a hardened patent-remedy professional could bring to bear. Even more boldly, Weiss and his colleagues also decided that there was no reason why *their* Bayer Aspirin and the other drugs should be marketed only in the United States. As Leverkusen's intellectual property rights were under attack all over the Allied world, why shouldn't Sterling try to challenge them overseas too?

It all sounded fine in theory, but there was a problem. When all was said and done, Sterling Products was still just a patent medicine business. Its bosses had no idea how to go about making the sophisticated drugs they now owned and the vast modern Rensselaer plant was completely mystifying to them. The German managers who had once run its production lines had all been sacked, interned or deported, and perhaps understandably hadn't left behind a handy set of instructions showing how everything worked. Nor was there any help from the Bayer patents. Sterling's executives found them as confusing as Mr Justice Joyce had so many years earlier. The documents truly were 'erroneous and misleading . . . by accident, error or design so framed as to obscure the subject as much as possible'. There was still a large amount of product stockpiled at Rensselaer, but it wouldn't last for ever. Unless Sterling Products found a way to get the plant working, its $5.3 million investment could turn out to be one of the worst bargains in history.

There was, of course, one possible solution, but it was not an attractive one. Sterling Products could try to get help from Rensselaer's former owners, Farbenfabriken Bayer. As these were the very people whose interests Sterling was now plotting to undermine, the 'alien' company that the American government had been so keen to get shot of, it didn't look a promising route to follow. There was a strong chance that the Germans wouldn't even talk to them. Fortunately there was one former Bayer executive still with the company, its export manager, Ernst Möller. He had kept his head down during the wartime farragos that had overwhelmed the other senior managers and had somehow clung on to his job. Seeing an opportunity to establish himself in the Sterling hierarchy and pay off a few debts to his old employers in the process, he offered himself as a middleman and pressed Weiss to do the only thing open to him. Go to Germany and cut a deal.

Had Carl Duisberg known what they were planning he would probably have enjoyed the delicious irony of Sterling's dilemma, but right then he had other things on his mind. At the end of April 1919, the German government's delegation arrived at Versailles to begin negotiating a peace treaty that would formally end the war. The hopes of the nation had gone with them. Many Germans hoped that the Allies would stick to the principles set out by President Woodrow Wilson in January 1918. His 'Fourteen Points' had covered such matters as the restoration of conquered territory, independence for nations such as Poland and removal of trade barriers, but otherwise seemed to offer a reassuring blueprint for a 'just peace'. In fact it all looked so reasonable that there was even optimism that Germany might get back some of its own lost assets. Among the delegates was a representative of the embryonic IG Farben combine, Carl Bosch of BASF, who hoped to secure the return of thousands of the group's patents, products, factories and trademarks still in Allied hands. Like everyone else with such interests overseas, Duisberg and the management of Farbenfabriken Bayer were on tenterhooks as to what Bosch might achieve.

It didn't take long for disillusionment to set in. It seemed the German delegation had not been invited to Versailles as equal negotiating partners, but as inferiors who were to be given a set of demands to which they were expected to agree. Shortly after their arrival they were presented with the first draft of the peace treaty – a document that they had had no hand in drawing up. It contained no mention of any of their seized assets or indeed of anything approximating to a 'just peace'. President Wilson, perhaps weakened by the long-term effects of the flu, had clearly lost his battle to persuade the Allies of the importance of allowing the Germans some shreds of dignity. The other victors, particularly the French, seemed more set on making the vanquished pay for their losses and preventing them from ever becoming a threat again.

The final document, signed by the demoralized and browbeaten losers on 28 June, was savagely retributive. Germany lost about 13 per cent of its territory, including all its overseas colonies. Its army was reduced to a rump of 100,000 men, and most of its and the navy's weapons were confiscated. The Rhineland was to be occupied and permanently demilitarized and France was given control of the industrially rich Saar region. To cap it all, the financial reparations demanded were almost impossibly high – sums far in excess of Germany's capacity to pay.

For Farbenfabriken Bayer and IG Farben the treaty was a disaster. As a grim-faced Carl Bosch explained on his return, one of its most damaging reparations conditions was that Germany had to immediately surrender 50 per cent of its chemical, dyes and pharmaceutical stocks. Even worse, for the next five years the Allies would be able to buy a quarter of any of those products at prices well below market rates. Not one seized trademark or product had been handed back, nor did it look as though they ever would. Short of a miracle, Carl Duisberg would have to accept the unthinkable. In much of the Allied world, Bayer's once jealously guarded aspirin monopoly was gone for good.

In such a climate, it is not surprising that Ernst Möller had found it difficult to bring his old and new employers together. His first letters to Leverkusen had received only a curt acknowledgement in reply and it was not until a frustrated William Weiss insisted on travelling to Europe himself that matters were brought to a head. In September 1919, at a small hotel in Baden Baden, Duisberg and Weiss came face to face.

They didn't hit it off. Duisberg obviously felt that Weiss was little more than a carpet-bagging opportunist who had managed to acquire something that rightly belonged in Germany. Only a few weeks earlier the Farbenfabriken had had to hand over half of Leverkusen's stock of drugs and other chemicals to the Allies. It had been a dreadful blow and now Duisberg wanted to redress the balance a little. At the very least he wanted his American aspirin business back. But this was one thing that Weiss absolutely refused to do. He knew the value of what he had bought and there was no way he was going to relinquish it now. He needed advice on the Rensselaer plant but he wasn't prepared to sacrifice the aspirin gold mine to get it.

Not surprisingly, the talks were inconclusive. Both sides left the meeting thinking that they had the beginnings of a deal (in which the outlines of a possible working partnership in Latin America was meant to have been the starting point for a wider relationship elsewhere), but it fell apart once Weiss went back to New York and Duisberg had retreated in a cold fury to Leverkusen. The problem was Duisberg's reluctance to leave the aspirin problem unresolved. He had decided that unless Weiss made a concession on that there could be no agreement.

A few months later Weiss and Möller returned to Europe to try again, this time attending a series of meetings at Farbenfabriken's Great Hall at Leverkusen. Again a deal was thrashed out. Again it fell apart. At one point Duisberg gave vent to his anger over Sterling's presumption that it had any right to use his company's good name.

> Everywhere in the whole world, except in the United States, people will say that we are the true Bayer. Laws say what they want; this situation contradicts global morality. They can't use our prestige for their advantage . . . No money is good enough.

His remarks set the tone for three years of sour transatlantic horse-trading. An interim agreement was reached on 28 October 1920, but it focused only on the aspirin business in Latin American territories. Sterling got an exclusive fifty-year licence to Farbenfabriken Bayer's Latin American trademarks in return for which the Germans would take 75 per cent of the profits. The Bayer Company of New York (as it was now known) would sell nothing else in Latin America under that name without Leverkusen's permission.

The rest took longer to put together. As the talks had gone on, Weiss had become more ambitious, wanting US sales rights to every Farbenfabriken product except dyestuffs. Think of what could be achieved, he had argued, with Leverkusen's technical know-how and Sterling's marketing expertise. Duisberg was unmoved. Much as he still hankered after the American aspirin business, and at least a share in the profits from the other lost products, it was preposterous to suggest that Bayer should hand over rights to everything else that could be possibly be made at Rensselaer to get them.

But finally, on 9 April 1923, just when it had begun to seem impossible that they would ever agree, the two sides reached a deal that divided the world between them.

Sterling's Winthrop Chemical Company would make all the Farbenfabriken products in America (but only America) and would get the help it needed from Leverkusen on how to run Rensselaer. The Germans would get half the profits in return. Sterling's Bayer Company would have the exclusive right to sell those products in the United States, Britain, Australia, Canada and South Africa. It could also continue to sell Bayer Aspirin in Latin America on the same 75/25 profit split as before, with the German company getting the lion's share. Everywhere else in the world would be Farbenfabriken Bayer's territory.

It was a complicated agreement and an important one. Sterling got the rights to sell some of the finest chemical and pharmaceutical products in some of the most important markets in the world. Not a bad return for $5.3 million after all.* For its part, Farbenfabriken Bayer had got a toehold back in the United States (through a share in the profits rather than through ownership) and had managed to keep an aggressive competitor at bay elsewhere.

More significant, though, was what the deal left out. Sterling Products still had the exclusive rights to Bayer Aspirin in America. Duisberg's last attempt to wrest back US ownership of his favourite wonderdrug had foundered. The miracle had failed to materialize. The task of steering aspirin's future through the land of golden opportunity and huge profits would stay in other hands.

It wouldn't be the end of the matter; the two companies had too uneasy a relationship for that. But at least they had established a short-term way of working together. This was just as well because they both now had other more pressing problems on their plate.

Duisberg and Farbenfabriken Bayer were about to become enmeshed in the financial insanity that overwhelmed Germany between 1923 and 1924, when the Weimar government's attempts to stimulate the economy (and meet its reparations bill) by issuing a flood of cheap money caused massive hyperinflation. During the brief but bizarre period that followed, when a humble tube of Bayer Aspirin tablets could change hands for almost a billion marks, it was impossible for any German business to do anything but hunker down and try to survive the crisis.

---

* Its only shortcoming was its necessary secrecy. Weiss knew that the American government wouldn't look too kindly on any commercial relationship with a company that only a few years before had been at the heart of the enemy's economic programme.

Sterling Products had its own concerns – more prosaic, but in the long term just as significant. Weiss had doggedly hung on to Sterling's ownership of the American Bayer Aspirin brand because he was convinced that it would one day pay handsome dividends. But almost before he had a chance to realize the vast profits of his dreams, one of the main reasons for buying it fell apart. Sterling wasn't the only American company with designs on aspirin and suddenly the competition had got a lot fiercer.

One of the Bayer Company's last acts before its seizure by the APC in early 1918 had been to file a test suit against the United Drug Company of Boston for infringement of its 'aspirin' trademark. United Drug was a wholesaler that had bought aspirin powder from Bayer in order to make it into tablets. These it had sold under its own name as 'Aspirin. 5 grains U.D. Co.'. The arrangement had ended in 1915 when Bayer began making its own logo-stamped pills because it wanted to make sure that the American public knew who produced the drug. However, when the Bayer patent expired in 1917, United was one of several US companies to take advantage, picking up supplies of ASA from new producers like Monsanto and marking its tablets in much the same way as it had earlier. So Bayer took it to court to protect what it saw as its property. Although the case was quashed on the outbreak of war and the subsequent auction of the Bayer Company, Sterling had immediately counter-claimed to re-establish the trademark rights, citing United Drug as plaintiff once more.

When the matter came to court on 17 May 1920, the issue before Judge Learned Hand was therefore whether aspirin was either a generic term for ASA or a brand of ASA. If he decided it was generic, then anyone in America who made ASA could call it aspirin. If not, then the only company that could sell it under that name in the United States was Sterling's Bayer Company. Rivals could call the drug besprin or zossprin or any other prin they liked, but not aspirin.

The case ran on for six days and in its way was as significant as the one Bayer had fought, and lost, over its patent in Britain back in 1905. After sitting through the usual interminable hours of arguments from both sides, Judge Hand came back with a complicated compromise solution. Aspirin, he said, was well known to wholesalers and pharmacists as a Bayer product. Therefore for sales of ASA to them, only Sterling could call the product 'aspirin'. But to the general public, the word 'aspirin' had passed into common usage and they had the right to buy the drug under that name from anyone who made it. For consumers therefore, aspirin was a generic term.

As the consumers were the only ones who mattered, this instantly deprived Sterling of one of its key advantages. In America, to all intents and purposes, Bayer Aspirin was now just an aspirin brand like anyone else's, even if it was still, for the moment, the most famous of those brands. It was a big blow to Weiss who had counted the exclusivity of the trademark as the drug's potentially most profitable asset. Its loss was one of the main reasons he had been so keen to get compensatory rights to Bayer's markets in the rest of the world and why, during all the time their protracted partnership talks were going on, Sterling and Leverkusen had been simultaneously skirmishing in courts throughout the world over who owned what trademarks and brands and where. In Canada, for example, Sterling won a case that meant that Bayer Aspirin was the only ASA that could be called aspirin.* In Mexico, on the other hand, the matter was irresolvable and Leverkusen and Sterling would battle for years over who owned the name.

But the most immediate result of the United Drug case was an explosion in competition in America that Weiss hadn't bargained for. Within ten years there would be several hundred different brands of aspirin on sale in US drugstores, all of them with the same active ingredient. It would open a chapter of commercial warfare almost unprecedented in modern times. With nothing chemically to choose between the products, the aspirin rivals would have to find other ways of convincing consumers that their tablets were better than the rest, dreaming up novel attributes and qualities that might appeal to the public. It

---

* Even now only Bayer Aspirin can be officially sold as aspirin in Canada. All other brands have to be called something else. It's just one of the complicated set of trademark rules and provisions that emerged in different countries around the world during and after the 1914–18 war, many of which still apply today. In Germany and seventy other countries Bayer still officially hold the aspirin trademark. In the United States, Britain and other places too numerous to mention, aspirin was ruled a generic term that anyone could use. It ceased to matter over time, of course, because wherever consumers buy the drug these days, they know aspirin by its generic name, regardless of which brand they actually purchase. However, Bayer still insists on putting a registered trademark sign alongside the name in all of its literature.

wouldn't take a genius to work out that in those circumstances advertising would be key, but, of course, it would take a genius to find the words and images that would make the difference. As it turned out, many of the new aspirin makers would make a pretty good fist of it, particularly those with enough money to spend down on Madison Avenue, but what they really all needed was a salesman with the golden touch.

Unfortunately, one of the very best of them was already accounted for.

George Davies was enjoying life. For the first time since he had left New Zealand (from where he had been forced to make a hurried departure to avoid the creditors closing in on his failed clothing business) he had a found a line of work to which his brash, if sometimes eccentric, selling skills were perfectly suited. He had at his disposal a highly useful product that was ripe for exploitation and he had employers who were willing to take risks.

The Nicholas brothers had done well in bringing their own new brand of aspirin on to the Australian market at a difficult time and had managed to sell enough Aspro to keep ahead of their creditors. But although George and Alfred had achieved much, the drug had not yet proved to be the surefire success that they had hoped for. Then George Davies swept into their lives and everything changed. Impressed by his enthusiasm, they had agreed to give him his head and had sent him away to think up a new marketing plan. When he came back and said that they should give away £2,000-worth of their product, they were a little nonplussed. It was a large sum for a struggling company, after all. But they decided to give it a try.

The idea was simple, Davies told them. Pick a target area (Queensland was chosen for the first trial run), put Aspro into threepenny packets and then hand them out for free to the public. Meanwhile, back up the promotion with as much local advertising as you can afford. It is arguable whether this was the first time that massive amounts of free samples were given away in support of a marketing campaign, but it was certainly the first time it had been tried with a reputable medicine. As Davies would explain in an accompanying advertisement, one of the first to be written in his own uniquely quirky style,

CHEMISTS GIVE AWAY £2,000 CASH TODAY

'Cast your bread upon the waters and it will return after many days.'
A phrase like that has lived 1,000 years.
Put into commercial language, today, it reads 'If you have got something good to sell, pay the public to try it – prove it good,' after that it will sell itself.

The strategy was successful. Just as Davies had promised, sales of Aspro in Queensland shot up as people tried it, found it good, and paid for the next packet in cash. So the company attempted it again, this time in Victoria, and again it was successful. Then came New South Wales and the Northern Territories. In each place the same tactic of free samples followed by aggressive advertising was tried and in each place subsequent sales of Aspro showed a marked improvement.

Of course, much of the success of these campaigns was due to the newspaper advertising copy that supported them, and that was where George Davies's real flair began to show itself. Like all good salesmen, he had very quickly convinced himself, totally and utterly, of the efficacy of the product. An Aspro tablet wasn't merely a few grains of aspirin. It was a uniquely splendid form of the drug – pure, safe and effective – that could do all sorts of wonderful things. It was his duty, nay his pleasure, to let consumers into the secret. His advertisements were a reflection of this passion, often scaling heights of corny hyperbole that made the brand a much talked about and much desired commodity. One began:

An Aspro tablet now and then
Builds up and soothes the greatest men . . .

It ran alongside a picture of Lloyd George, who, of course, wouldn't have known an Aspro tablet from an indigestion pill. Another advertisement (betraying Davies's penchant for arbitrarily enlisting the endorsement of famous historical figures) began:

Hats off to the great Abe Lincoln, because he was a stickler for the truth. As such he would recognize a wonderful discovery like Aspro in a minute.

At other times his copy took the form of concise melodramas about the detective who made a vital arrest only because Aspro had kept his mind clear at the critical moment, or the nurse who had been in agony for fifteen years until Aspro cured her aches and pains. And alongside each would be a list of ailments for which Aspro was supposedly efficacious – a catalogue of disorders of which even the infamous Lydia E. Pinkham would have been proud: headaches, colds and rheumatism, of course, but also sciatica, gout, upset stomach, anxiety, insomnia, 'pains peculiar to women', nerve shock, irritability, and much else. It didn't matter that there was little or no available scientific justification for many of these claims, several of which were so vague as to be irrefutable. Davies had convinced himself that they could be true and that was enough.* Not surprisingly, when the flu pandemic hit Australia in 1919 and the Australian government put Aspro on the necessary commodity list, Davies's delight knew no bounds, and the ads started carrying little drawings of the Federal Parliament building as well.

> This is the first time on record that a responsible Minister of the Government of any country had proclaimed a Popular Medicine a necessary commodity in the interests of the community. This action speaks louder than words of the tremendous importance of Aspro as a medicine for the Human Race . . . There's nothing else like it.

Of course there were other things like it – other aspirin brands, for a start – but Davies's copy left the consumer in no doubt that Aspro was head and shoulders above any of its rivals. And although these advertisements look crude to modern eyes they were written in such a way as to stay just inside the acceptable standards of the time. Australia in those days had few of the constraints on medical advertising that exist today, and even though some doctors and pharmacists demurred at the often outrageous claims that were being made for Aspro, there was little they could do to stop them. It helped, of course, that aspirin was indeed a remarkable drug. Unlike most patent medicines, it really did

* Bearing in mind the many remarkable things that scientists now know that aspirin can do, this list doesn't look quite so far-fetched today, but at the time Davies was stretching the drug's properties to a quite extraordinary degree.

relieve pain, bring temperatures down and make people feel better – just not quite to the extent that Davies's advertisements claimed.

The drug's unarguable efficacy was probably what calmed the consciences of George and Alfred Nicholas, who would never have dared to run campaigns like this on their own. The hugely increased sales helped too. Aspro profits began to mushroom. Soon the company had enough money to sell its high street pharmacy headquarters and move into larger premises elsewhere in Melbourne. It celebrated by changing its name to Nicholas Proprietary Limited and by upgrading its creaky old production process. There were soon other moves afoot as well. With a lead firmly established in the Australian market, George and Alfred Nicholas wanted to expand their horizons overseas. They opened up a factory in Wellington, New Zealand, which did well, and then looked farther afield. It was time to start taking on the pharmaceutical giants of Europe and America in their own backyard.

The war had changed aspirin production in Britain beyond all recognition. Gone were the days when it was almost exclusively an imported product, brought in by one or other of the German companies. From the moment that the government had renounced Bayer's trademark rights to the aspirin name, British chemical and pharmaceutical manufacturers had added it to their product lists and some, after a few hiccups, had managed to make quite a good business out of it. The tiny number of pre-war British ASA producers, led by Burroughs Wellcome, were joined by new aspirin makers such as Thomas Kerfoot Ltd, a Lancashire business, whose Salasprin brand had finally found favour with the British Army, Genatosan of Loughborough, which made Genasprin, and the Boots Company of Nottingham which made its own generic version of the drug for sale in its nationwide chain of high street chemist stores. Other brands, like Empirin, Acetysal and Aletodin, also began to make an appearance. As had happened everywhere else in the industrialized world, the flu pandemic had increased demand for all these aspirin products and by 1920 the drug was the biggest-selling analgesic in Britain.

But none of the new brands had yet managed to build a clear lead over its rivals, and when Farbenfabriken Bayer re-entered the British market after the war the picture became even more distorted. Its American antagonist and soon-to-be-partner, Sterling Products, had bought the confiscated UK rights to the Bayer Cross trademark and the assets of the Bayer Company Ltd in 1919 from the Board of Trade. As a result, it claimed exclusive ownership of the Bayer Aspirin brand in Britain. However, Carl Duisberg disputed the legitimacy of this deal and licensed other UK importers to sell Leverkusen's version of the drug. For two years both companies had sold their own product in Britain, each confusingly bearing the Bayer Cross logo and each called Bayer Aspirin. The matter was finally resolved as part of the agreement that Weiss and Duisberg struck in 1923, but in the meantime bewildered pharmaceutical wholesalers had become unsure which one to stock. Neither brand had therefore been able to re-establish Bayer's pre-war dominant position. It meant that although Britain had become a nation of avid aspirin consumers, the public were in the habit of buying whichever product came to hand. Consequently, the market was wide open to the first producer, domestic or foreign, which could put together a coherent sales and advertising strategy and make its drug stand out from the rest. This was what the Nicholas brothers now sought to do.

Their first attempt was a disaster that nearly destroyed the business. In 1924, the company sent over its chief financial adviser, George Garcia, to take a look around and oversee a test-bed promotion of Aspro at the Empire Exhibition in Wembley. Garcia spent a few weeks travelling the country and then cabled headquarters saying that because the prospects appeared to be so good it didn't make any sense going on to try to crack the difficult US market. Nicholas Proprietary should put all its effort into Britain. So while Melbourne began making arrangements to ship over the first large batch of stock, Garcia went in search of an advertising agency to prepare a campaign aimed at two target areas, Lancashire and Yorkshire, that were deemed to be especially under-exploited.

But a few months into the launch it became clear that it had all gone horribly wrong. In deference to the supposedly more refined sensibilities of the British public (and the genuine constraints posed by stricter UK trading laws) the full-blown Aspro marketing campaign had been toned down. The idea of giving away free samples was followed, but George Davies, to his disgust, had been left at home and the task of drawing up the necessary Aspro advertisements had been passed to a London agency, which underplayed the product. What's more, their efforts coincided with a deep economic depression in Britain. Unemployment was high and there was little spare cash around. Consumers redeemed their newspaper coupons for free packets of Aspro all right, but didn't bother to buy the product a second time. The campaign was tried again and again, but with the same lamentable results. Costs began outrunning sales on a scale of seven to one.

Alarmed by the reports coming from London, Alfred Nicholas decided to go and investigate. He was appalled by what he found. The British enterprise had lost £200,000 and was going nowhere. Unless something was done, losses on that scale could completely cripple the whole firm. The only sensible thing to do was to pull out of Britain altogether.

Loath to accept total defeat, however, Alfred cabled his brother George back in Melbourne and asked him for his views. The answer was unequivocal. Give it one last try, in a smaller area, and hit it with everything you can. George Davies is on his way.

To the irrepressible Davies this was his greatest challenge yet. If the Poms hadn't fallen in love with Aspro it was only because no one had got around to telling them how good it was. It was time for the 'gentlemen' in the company's London advertising agency to stand aside and let the master work his magic. He arrived in Britain to hear that Alfred had decided to target the unsuspecting citizens of Hull in Yorkshire. Davies found a typewriter, rolled up his sleeves and started to write.

The advertisements were masterpieces of salesmanship – and invention. Once again, Davies used his favourite tactic of running copy alongside pictures of the great men of the day, carefully including a cross-section of political leaders to suit all tastes. Ramsay Macdonald, Stanley Baldwin, Austen Chamberlain and Lloyd George were all thus unknowingly recruited to add authority to his message.* Even the King was quoted in praise of products from the empire, although of course he had never specifically referred to Aspro as such. But Davies wasn't content to restrict himself to celebrity endorsement. He drew his inspiration from wherever he could find it.

* Of course, whenever he tried this, some of the 'great men' involved would write and complain that their names and images had been taken in vain, but by then it was too late. Davies would promise (and even sometimes deliver) a small footnote apology on the next advertisement, but by then of course, he had also moved on to another unsuspecting endorsee. He was never short of potential candidates to choose from.

> The science of atomic energy is now more than mere anticipation – scientists tell us we are on the verge of a great discovery. It is found that there is enough atomic energy in the little finger to run all the trains in England several minutes if it could be harnessed. Truly this is progress. Similar happenings have been demonstrated in the medical world by Aspro.

Davies also struck a particularly rich theme when he told tales of how ordinary people in extraordinary situations had found the product uniquely beneficial. One of the most entertaining advertisements from the time was headlined: 'Aspro. Message from the Bottom of the Sea.' Under this was a dramatic image of a surfacing submarine, alongside a portrait of a sailor, one Able Seaman Jevons, in full Royal Navy uniform. The accompanying copy ran:

> H M Submarine L. 71 DEVONPORT
> Dear Sirs – I have served in Submarines for 7 years and I can say that life in Submarines is a very nerve trying ordeal, and at the actual time of writing we are lying at the bottom of the Channel.
>
> In conditions like this we are always on watch, and the atmosphere at times becomes extremely thick and nauseating, which, after hours of breathing, gives me a terrific headache. After trying several well-known remedies I find that the only really permanent one is your 'ASPRO' Tablets, and all the crew now take them as we find that they give instant relief.
>
> This testimonial is quite unsolicited and you may use it to the best of your advantage.

This last statement was true of course: the testimonial was totally unsolicited – but only because Able Seaman Jevons didn't exist. He was just one of George Davies's cast of imaginary characters, snatched out of thin air to help sell the company's pills.

The impact of all this on readers of the *Hull Daily Mail*, where most of these creations were placed, can only be imagined, but they must have been rather startled. It wasn't that they were unused to advertisements *per se* – the paper was regularly full of blandishments for such things as Army Club cigarettes, Neaves's Baby Food (For the Infant Prodigy!), Crawston's Corsets, and Dr Cassell's Liver Pills – but the Aspro ads were so much more flamboyant. They were also huge, taking up a third or more of a page, and as the weeks went by, and George

Davies's copy became more ebullient, and the apparently unsolicited testimonials from grateful patients more spectacular, they would have been impossible to ignore.

'Thank you for your wonderful achievement. They are a Godsend to suffering humanity,' endorsed one such sufferer, whose ten years of violent headaches had apparently been cured at a stroke by Aspro tablets. 'The most wonderful pick me-up. I shall insist on taking them always,' wrote another under a headline that screamed, 'From 12 million to 240 million tablets in four years. A WORLD RECORD!'

His inventiveness was given an extra spur during the winter of 1926–7 when another flu epidemic broke out in Britain. It was considerably milder than the one at the end of the war, and would kill only a handful of people, but memories of that earlier disaster clearly still lingered in the public's imagination. Davies exploited those fears with all the zest he could muster. 'ASPRO SMASHES THE FLU IN ONE NIGHT – POSITIVE PROOF,' he wrote, over yet another endorsement of the drug's 'miraculous' properties, this time featuring a businessman whose entire workforce had been laid low with the flu until they started taking Aspro.

Hull was bombarded with this material for week after week, and inevitably it began to pay off. Sales of Aspro rose dramatically and the campaign was extended to the nearby city of Leeds and then to the rest of Yorkshire. By the end of the year it was running nationally. The company opened a new British headquarters and factory in Slough and soon Aspro was the best-selling aspirin brand in the country.

Understandably, Alfred Nicholas was delighted. His last throw of the dice had paid off. He returned to Australia with a triumphant George Davies in tow. Days after their return they began plotting their next assault – on markets in continental Europe and throughout South East Asia. But the success of the campaign had not gone unnoticed by rival manufacturers in Britain. These responded with equally ambitious campaigns of their own (particularly Genatosan, whose bold Genasprin advertisements became a particular feature in the popular *Daily Sketch*), and although it would be some time before they caught up with Aspro, they cumulatively pushed the volume of aspirin sales ever higher across Britain, doubling the market in less than five years. Davies, it seemed, had made Britain a nation of pill-poppers, and given Bayer a nasty commercial headache in the process.

One market remained unchallenged, though – the biggest of them all. Nicholas Proprietary dipped its toe in and out of the United States several times but was unable to make the breakthrough that it managed elsewhere. The country was too big, there were too many competitors, and in any case it was Sterling's home turf. As William E. Weiss would prove, George Davies wasn't the only one with original ideas.

Sterling Products' founders had promised to spend lavishly on aspirin advertising after it acquired the US rights to Bayer, but for the first few years after the war they found that they didn't have to. They issued ads a-plenty, of course, but they were far more modest than those Aspro was throwing at the British market. Most consisted of a simple picture of a Bayer Aspirin bottle with the same standard list of ailments – colds, headache, rheumatism, toothache, lumbago and neuralgia – that every aspirin producer stuck on its packaging.*

This reticence was partly caused by an innate reluctance to upset the American medical authorities further, which Sterling had inherited from the brand's German owners and found hard to shake off, and partly by the very diverse nature of the competition it faced. So many new aspirin brands had appeared on the US market in the aftermath of the United Drug trademark case that Bayer, perversely, benefited from the confusion that this caused consumers. The new makes all jostled for position, but in the meantime the power of the Bayer name paid dividends. Consumers knew the product from the days when it had been the only aspirin available and many stuck with it, even though it was one of the most expensive versions of the drug available. Bayer Aspirin remained loftily above the rest as the market leader.

This state of affairs couldn't last for ever, though, and eventually Sterling's rivals had begun closing the gap. As there was little or nothing to choose between any of their products in terms of chemical composition or even their effectiveness, Bayer's rivals had had to dream up other ways of standing out from the herd. This led to ever more imaginative marketing campaigns and increasingly inventive advertising. Thus Burton's Aspirin was sold on the basis that it 'doesn't nauseate', Molloy's Aspirin was 'stronger than all the rest', Cal-Aspirin was better

* This reticence hadn't prevented Sterling from throwing money at advertising in other markets, particularly in Latin America.

because of its added calcium (which made little actual difference to the drug), St Joseph Aspirin was 'made with the family in mind', Cafasprin was enhanced 'by vitalizing caffeine', and so on.

Other products were sold on the basis that they were soluble (although those that claimed to be, weren't; a proper formula for soluble aspirin still lay some time in the future) or in powder form that could be added to water and taken as a sparkling drink. One of the first of these effervescent products was Alka-Seltzer, from Miles Laboratories of Indiana. According to company legend, its President, A. H. Beardsley, stumbled on the idea when he paid a visit to a local company during a flu outbreak in 1927 and found that employees were all being kept healthy by a daily drink of aspirin and bicarbonate of soda. Beardsley was so impressed that he got his head chemist to make up a similar compound, which later became the well-known brand.

Eventually the advertising supporting all these new products began to have an effect, and Sterling was forced into an aggressive and innovative response. It counter-attacked through two new media. The first of these was the roadside billboard. The 1920s and 1930s saw an explosion in automobile ownership across the United States, accompanied by a marked growth in road-building programmes. America's new love affair with the car took its citizens on increasingly lengthy journeys over these highways and it didn't take advertisers long to realize that drivers could be a (literally) sitting target for bold, brash ads. Soon every major road had its plethora of hoardings, and Sterling's Bayer Aspirin became one of the most prominently featured names.

But it was through radio that the brand made its biggest mark. The idea of broadcasting advertisements, which seems so normal today, was a bold new concept in the 1920s. Radio operators had discovered that they could make money by selling their airtime to makers of cosmetics, cars, beer and other mass market products, which led in turn to sponsored programmes. For relatively little investment (radio advertising was often much cheaper than a big newspaper campaign), these companies were able to target increasingly large and often previously unreachable audiences.

With an old patent medicine huckster's instinct for a captive audience, Weiss saw the huge potential in radio and threw money at it. Bayer Aspirin sponsored news programmes, broadcasts of sporting events, variety and musical shows and much else. The words, 'This programme is brought to you by Bayer Aspirin, the Genuine Aspirin,' became familiar in homes across the country, with the shows themselves interspersed with announcements to the effect that the drug did not affect the heart, or giving dire warnings of the perils of 'counterfeit' non-Bayer Aspirin products. By the early 1930s Sterling was spending up to half a million dollars on radio commercials alone.

The often exaggerated claims made by Sterling and its rivals did not escape the attention of the authorities, of course, and several producers of leading brands fell foul of the Federal Trade Commission's rules on false advertising, mostly for saying that their particular make of aspirin was different from others. But the admonishments rarely amounted to more than a slap on the wrist. One of the unfortunate consequences of the battles over America's food and drug legislation in the years before the war was that the FTC's remit covered only advertising infringements that damaged competition and had no effect on claims relating to the supposed medical benefit of the drugs, whereas the Food and Drug Administration (the successor to Harvey Wiley's Bureau of Chemistry), which oversaw the composition and purity of drugs, had no remit to regulate medical advertising. This topsy-turvy situation would eventually be resolved by legislation some years later but in the meantime it left the aspirin makers a large loophole. Moreover, the American medical establishment and the official pharmacists' associations could do little about it because now aspirin was being sold as an over-the-counter product and therefore lay outside the purlieu of official prescription drugs.

And so the advertising battle went on with increasing ferocity, with millions of dollars being spent by America's aspirin producers on advertisements that made much of the cosmetic, if essentially non-existent, differences between one brand and another. Occasionally one of the less popular products would go under, to be replaced by a new brand, but most of the major names, like Bayer Aspirin, managed to improve or at least cling on to their market share. In the meantime, of course, total aspirin sales were on the rise, as the little white pills became an ever more familiar fixture in bathroom cabinets, desk drawers and handbags. This was the nature of effective advertising, its ability to convince a consumer to buy a product that they didn't necessarily need. As the English humorist Jerome K. Jerome had written several years earlier, 'It is a most distressing thing, but I never saw a patent medicine advertisement without being impelled to the conclusion that I am suffering from the particular disease therein, dealt with in its most virulent form.' Many aspirin consumers were now buying the tablets simply because the ads told them that they should.

The drug's burgeoning popularity and public profile meant that aspirin soon became a cultural icon. It was seen as a truly modernist product, a scientific marvel that anyone could obtain for a few pennies. It wasn't for nothing that this era was dubbed 'the aspirin age' by the writer Jose Ortega y Gasset. In 1930 he proclaimed (a little optimistically) that,

> the man in the street today lives more easily, more comfortably and more safely than the potentate in the past. It matters little to him that he is no richer than his neighbour if the world around him is rich enough to provide him with roads, railroads, hotels, a telegraph system, physical well-being and aspirin.

Franz Kafka, meanwhile, extolled its virtues to his girlfriend; Enrico Caruso famously demanded aspirin from his manager before he would go on stage and George Orwell, in *The Road to Wigan Pier*, tried to explain why the British working classes preferred the affordable luxury of stimulants like tea and aspirin to the more obvious nutritional benefits of wholesome brown bread. Aspirin even made its appearance on board Graham Greene's *Stamboul Train* and got a George Davies-like endorsement in Edgar Wallace's *Door with Seven Locks*: 'The hammering gradually lost power over Sybil, the roaring in her ears faded to a soft humming and the fog suddenly lifted from her memory.' The fortunate Sybil had, of course, just taken one of the famous little white pills.

But with this fame came problems. Aspirin was a hot product, huge sums of money were being made from it, and it became a target for the criminal world. Counterfeiting, which Bayer had struggled against before the war, resurfaced as a threat. Alongside the legitimate aspirin producers were hundreds of unlicensed back-street operations turning out contaminated and sub-standard versions of the drug. In Germany, Bayer had to launch a national newspaper campaign threatening to sue any pharmacist found with contraband product. The situation became so bad in Norway that the nation's Pharmaceutical Society began calling for life imprisonment for criminal counterfeiters.

Some criminals took the more direct route. In early 1927, a group dubbed the Aspirin Ring Gang broke into a Sterling Products warehouse in Hudson Street, New York, and got away with $92,000 worth of Bayer Aspirin (over a million tablets). They were caught only because Moe Stratmore, the leader of the eight-strong gang, was gunned down by police during a unrelated bank hold-up in Hoboken, New Jersey, and one of his associates, Eugene Steiner, who had been responsible for storing the stolen drugs at a garage in Brooklyn, turned State's evidence. Even then the case almost fell apart. The other defendants raised $10,000 dollars between them in an attempt to 'fix' the jury and arranged for Steiner's wife and children to be kidnapped and held hostage to stop him giving evidence. The District Attorney, Charles J. Dodd, was only able to get convictions in January 1928 after Mrs Steiner had been tracked down and rescued, and the charges were downgraded from theft to 'fencing'.

Increasing use of the drug brought medical problems too. The mechanics of aspirin's action weren't clearly understood at the time (and wouldn't be for another forty or fifty years) but some doctors and chemists had long believed that it affected the circulatory system in some way. This ran parallel to the mistaken theory, first raised as a possible problem by Bayer's then chief chemist, Heinrich Dreser, back before aspirin's launch, that it could have a damaging effect on the heart. Producers of rival analgesics made much of these fears – hence the statement, often found in aspirin advertisements from the time, that the product did not damage the heart. Of course, no one then knew the truth of the matter (the later discovery of its actual effect on the heart would come as a huge shock) or even how tolerant the body was in other ways to the drug, but it made some doctors sensitive to its misuse. Many of these concerns made their way into the pages of the *British Medical Journal* and the *Lancet* throughout the 1920s and 1930s.

For instance, in the *BMJ* of August 1920, a Dr R. Eccles Smith, of Barry, near Cardiff, wrote about a sailor in his care who had taken two and a half grains of aspirin, 'and within ten minutes he had severe headaches, profuse sweating and partial collapse'. The patient's body was also covered in a nasty rash and it was the second time he had suffered the same symptoms, having experienced similar problems with aspirin three months before. A Dr Ethelbert Hearn from Sheffield wrote back and said that he had experienced the same thing when taking aspirin himself and 'felt a bit seedy for a few days'. Another doctor, H. E. Davidson, wrote that he had a case where regular doses of aspirin were suspected as the cause of dreadful itching. The condition stopped when the patient stopped taking the drug. Could there be a connection?

Of course, the answers to some of these conundrums would take some time to find (and many would turn out to be diagnostic errors), but what they also reflected was a growing concern that some doctors, and indeed some of the public, felt about the widespread availability of aspirin and the potential for its misuse. Every few months the newspapers would contain stories of people who had tried to overdose on the drug (which, although harder than it sounds, was certainly possible) or who had even poisoned others. The details of one particularly tragic case in Lewes, Sussex, were widely reported in July 1929 and concerned a young mother who had dissolved 500 tablets of aspirin in a pint of water and then given her baby two dessertspoonfuls of the solution. The child died and the woman was subsequently committed as a 'criminal lunatic'. That same year, Sir Robert Thomas MP rose in the House of Commons and asked if the British government was aware of a proposal to install automatic aspirin-dispensing machines on street corners, and whether the uncontrolled sale of the drug was in the public interest. He was told the matter was under review. On 6 June 1931, at a meeting of the Medico-Legal Society in London to discuss dangerous narcotics, a Dr Gerald Stot raised the addictive potential of aspirin saying, 'It is not uncommon to see women eat aspirin out of their handbags as though it was so much sugar.'

On 14 September 1935, the *Lancet* sought to make sense of some of these concerns and, perhaps reflecting the significant recognition that aspirin had achieved, gave over its whole front page to the drug. To the relief of producers it came firmly down on their side, albeit with a barely concealed sniff at their sales techniques.

> Not the least of the contributions of chemistry to human well being is the gift of aspirin, now manufactured and swallowed by the ton. Its analgesic effects and its safety are so widely known that its use has long passed into

> the control of the general public, with much encouragement from the manufacturers and little guidance from the medical profession.

Although overdoses were rare, the article continued, they did occur and doctors should be aware of the possibility. However, it concluded: 'aspirin in ordinary doses appears to be singularly safe'.

It was a remarkable endorsement from one of the world's leading medical journals and must have been thoroughly enjoyed by the management at Leverkusen. Not that they needed any lessons from anyone about the drug's achievements. They had celebrated its success in their own way some two years before, when a huge illuminated 236-foot-diameter aspirin tablet, decorated with the Bayer Cross, was installed above the factory on the Rhine. When it was switched on, the lights could be seen for miles, a proud monument to the company's most famous product.

But things were changing at Leverkusen – and in Germany too. In 1938, the company published a lavishly printed booklet describing fifty years of Bayer pharmaceutical achievement. It proudly listed its many successes, aspirin among them, and concluded by describing how its modern technology was now 'always available to supply any doctor with Bayer preparations by the quickest route and in the shortest possible time'. The statement ran alongside a photograph of the company's new aeroplane, flying proudly down the Rhine over the giant Leverkusen complex. The wings of the aeroplane bore the famous Bayer Cross logo. Its tail fin was decorated with a large black swastika.

9

A MORAL COLLAPSE

A PECULIAR FEATURE of aspirin's story is the way it has sometimes exerted a distorting effect on the lives of those most closely associated with it – and through them on the momentous events of their times. Why fate should have marked it out for such a pivotal role in human affairs is hard to say. It's just an inanimate pharmaceutical product. By rights, outside the medical sphere, it should have had no more impact on our world than milk cartons or shampoo or any of the other useful (and useless) things with which we clutter up our lives. But aspirin is one of those rare commodities whose very existence seems to have influenced history, its invention provoking decisions and events that might not have otherwise occurred.

This is worth remarking upon because the next episodes of aspirin's story have their roots in the small but significant role that it played in the creation of a massive industrial cartel during the 1920s and 1930s – an organization that then became one of the buttresses of the most grotesque and brutal dictatorship the world has ever seen. The cartel was IG Farben and the dictatorship was that of Adolf Hitler. The line from aspirin runs through them both, and leads all the way on up to Auschwitz.

What a cruel irony, then, that among the many millions of people whose lives were destroyed by these events would be some of those most intimately associated with aspirin's invention and sale.

Farbenfabriken Bayer emerged out of the economic chaos that overwhelmed Germany in 1923–4 to find a world returning to some semblance of normality. A new Reich mark regulated by a new central Reich bank, stricter rules on government spending and some relaxation by the Allies of their reparations demands had all done their bit to burst the hyperinflation bubble and restore public confidence in the fragile German financial system.

For Carl Duisberg, who had managed to complete his wrangling with Sterling Products just before the economy went into meltdown, it was an ideal opportunity to return to his other grand design: his self-appointed task of bringing all Germany's main chemical manufacturers together into one organization. The semi-cartel that he had stitched together during the war had worked well, but in subsequent years the old competitive pressures had begun to reappear and now something more permanent was needed. However, in the meantime, he'd started to worry that the shock of totally amalgamating all the individual businesses might be too disruptive. Perhaps a better route would be a step-by-step process, starting with a jointly owned sales and investment company. A proper merger could come later.

To his surprise, the other members of the old cartel thought otherwise. Led by Carl Bosch of BASF, who was still seared by his experiences at Versailles, they had capriciously come to accept the view that Duisberg had expressed before the war, that only a full and immediate merger would let them re-establish their control of the world chemicals industry. These arguments now carried the day. On 15 September 1925, the six companies and their shareholders agreed to join together in a totally new combine. The deal was simple. While they would maintain all their old product brands and areas of expertise (Bayer Aspirin would still be Bayer Aspirin) they would all now be subsidiary divisions of a single organization, chaired by Carl Bosch. It would be known henceforth as the Interessengemeinschaft Farbenindustrie Aktiengesellschaft, although everyone would still call it IG Farben for short.

Losing the overall leadership of the organization was a dent in Duisberg's pride, but he retained a seat on the board (and much of his influence), and gradually the astonishing success of the huge new enterprise convinced him that he had been right all along. Within twelve months, IG Farben was capitalized at over a billion marks and was well on the way to becoming the largest company in the world. In the years that followed, through a labyrinthine network of subsidiaries, partnerships and holdings, it would go on to make and sell thousands of products, from drugs and explosives to dyes and synthetic petroleum. More importantly, as Duisberg had once predicted, by absorbing or crushing any effective competition, it was able to accrue unprecedented power over the prices it was able to charge and the markets it was able to control. Quite simply, it came to dominate the global chemical industry in a way never seen before or since.

Had this been the sum total of the IG's contribution to history, then perhaps future generations would have looked more kindly on its achievements and Duisberg's grand vision might have been more celebrated. Sadly, however, the IG is not generally remembered for its business acumen, but for its complicity in the crimes of the Third Reich.

The IG's association with the Nazis began on the evening of 20 February 1933, just a few hours after Duisberg switched on the giant Bayer Aspirin sign above the factory at Leverkusen. The National Socialists had come a long way since their failed putsch in Munich ten years before. Adolf Hitler had finally made it, via a spell in jail, a deepening political and economic crisis, and an easily manipulated President Hindenburg, to the office of Chancellor. Now the brown-shirted Sturmabteilung were strutting through Berlin, and the party apparatus was putting the squeeze on German industry.

A meeting had been arranged for that night because business leaders had realized they would have to deal with a man they had hitherto dismissed as a nuisance. None of them anticipated any problems, though. Beguiling shopkeepers and housewives was one thing, but if Hitler thought he could impress the cream of German industry he was in for a surprise. Top industrialists, including IG Farben's head of production, Georg von Schnitzler, filed into the room fully expecting to bring the Chancellor to heel.

In the event, it was they who were surprised. They barely managed to get a word in edgeways. Hitler harangued them about the threat posed by Bolshevism and the increasing decay in German society. The fatherland was in grave danger, he said. A parliamentary election was due on 5 March. If the National Socialists didn't win an outright majority, a civil war was inevitable. As the shocked businessmen absorbed this barely veiled threat, the Reichstag President, Hermann Goering, stepped up and smoothly delivered the sting. There was one way to avoid conflict, he told them. The party needed three million marks to win the elections. If the money could be found, then the Nazis would keep the peace and normal business activity could continue undisturbed.

It was blackmail, of course, but no less effective for that. Cornered and confused, some of the industrialists took out their chequebooks on the spot. Others promised to make a donation shortly. Although von Schintzler was unable to authorize any payment personally, he reported the matter to Carl Bosch, the IG Farben Chairman. A few days later Bosch reluctantly gave the Nazis 400,000 marks.

It was the beginning of a relationship that one day would see representatives from both organizations facing the judges at the Nuremberg War Crimes Tribunal. The first payments must have seemed a relatively small price to pay for industrial tranquillity but a week later the Reichstag was burnt down (ostensibly by a Communist agitator, most likely stage-managed by the Nazis) and Hitler had his scapegoats. In the outcry that followed, the cash-rich National Socialists won the elections and Hitler transferred all parliamentary powers to himself. Within a year his totalitarian dictatorship was firmly established.

Bosch was never the most enthusiastic supporter of the new regime. He thought Hitler's economic policies were risible and was embarrassed and confused by the Nazis' subsequent demonization of the Jews. Many of IG Farben's leading scientists were Jewish, as were four of its board members, and it was practically difficult, let alone morally repugnant, to force them to leave the company as the government's new race laws soon demanded. To Bosch's great credit he protected some Jewish employees by sending them abroad out of harm's way, and once even remonstrated with Hitler personally about the damaging effect of the policy on his business.* But he was hampered by the fact that he needed the Nazis' help to rescue IG Farben from a failing business venture. The combine had invested hundreds of millions of marks in a new process (invented by Bosch himself) to make synthetic petrol, but plummeting world oil prices had completely undermined the economic logic behind it. In desperation, he appealed to the government for aid and to his relief it agreed to buy all of the new fuel. But he'd struck a Faustian pact. By the time Bosch retired as Chairman in April 1935, the IG's ability to stand up to the Nazis had been fatally compromised.

* The meeting took place in May 1933 after the IG Farben boss was appointed to a new general business council. Bosch warned Hitler that the dismissal of Jewish scientists could set back German physics and chemistry by a hundred years. 'In that case,' the Führer is said to have replied, 'we'll work for a hundred years without physics and chemistry.'

A few weeks earlier an even more momentous passing had taken place. On 19 March 1935, Carl Duisberg died, aged seventy-three. More than anything else his death marked the transformation in the IG's affairs. A fierce patriot and a believer in strong leadership, Duisberg was none the less uncomfortable with the gangsterism of the Third Reich and it is probably just as well that he died before he could see where the Nazis were leading his country. Had he lived, it would have been hard for him to bear because he would have had to accept how his two lifelong obsessions – aspirin and the creation of a unified German chemical business – had been stepping-stones to calamity.

Aspirin, it must be remembered, had made Farbenfabriken Bayer a great deal of money and had been crucial in taking Duisberg to his pre-eminent position within the German chemical industry. Once established, he had been determined to mirror the power of American cartels by bringing Germany's leading chemical companies together (that he was briefly reluctant to pursue this strategy as aggressively as others does not alter the fact that he was the driving force behind it). But had aspirin never been invented, it is unlikely that Bayer would ever have become more than a medium-sized drug and dyestuffs company or that Duisberg would ever have been in a position to realize his ambitious plans. Germany's chemical companies would thus have remained in aloof competition and the grand cartel would never have been formed. How differently might things have turned out, then, had there been no IG Farben to prop up the Nazis?

Because prop them up it did. As history now knows, the IG went on to bankroll the Nazi Party, eventually to the tune of some 80 million marks. In sanctioning this support its executives, among them some of the most important and influential business leaders in the country, actively colluded with the Nazis' corruption of German politics and helped to create the moral vacuum that was the Third Reich. By the start of the Second World War, when the IG was at the zenith of its power, its industrial might was a critical factor in Hitler's hopes of realizing his political and military goals. The IG gave Nazi Germany the strategic materials – synthetic oil, rubber and nitrates – that it needed to fulfil its expansionist ambitions and prosecute a war. What's more, with a few honourable exceptions, the men running the combine did little or nothing to contest or disrupt these activities or deny its output to the Nazi military machine. Of course, it may have made little difference had they tried, but the the attempt wasn't made.

Nor were the combine's sins only those of omission, a necessary consequence of life under a police state. The IG acceded to the 'Aryanization' of its workforce – reluctantly and under pressure at first, but with increasing alacrity as the 1930s rolled on. And then, of course, it became directly complicit in the slave labour, plunder and genocide that the Nazis visited on the world during the Second World War. It produced, through its Degesch subsidiary, the Zyklon B gas that was used to murder millions of men, women and children in the Holocaust, it financed and managed some of the camps in which they were slaughtered, and profited in countless ways from the cruelty of the regime under which they met their fate.

No, Carl Duisberg would have found it all too much to bear.

The Leverkusen factory closed on the day of his funeral and crowds lined the street to see his cortège pass by. An obituary in *The Times* said: 'He may well come to be considered the most efficient and effective industrialist the world has yet known.' It was a fair judgement. Despite the unforeseen consequences of his great design, and the many failings he may have had as a person – autocratic, imperious and ruthless are epithets that even his friends applied to him – Carl Duisberg was undoubtedly a business visionary of extraordinary brilliance. In the dark days to come he would be sorely missed.

One of his lasting regrets was that his nemesis William E. Weiss still firmly controlled the US aspirin business. For all his other triumphs, Duisberg had never completely accepted this loss and though he had learnt to get along with the flamboyant American he might have enjoyed a moment of grim satisfaction had he stayed around to see what fate had in store for the boss of Sterling Products. IG Farben would not be the only organization to be tarnished by its association with the Nazis.

William Weiss had done well out of his deals with Leverkusen and did better still when his lawyers told him they also applied to the new IG Farben. He had to make some concessions, agreeing to give the cartel a 50 per cent stake in his Winthrop Chemical Company. However, in return Sterling Products got the sales rights to the new conglomerate's entire US business in pharmaceuticals, fertilizer and photographic products – a fantastic and unexpected bonus. The deals had to be kept confidential, of course, as they were in contravention of the terms under which the Alien Property Custodian had sold the American interests of the Bayer Company to Sterling Products in 1918. But as long as the money was rolling in, this didn't seem to matter much. In any case, Weiss had other things on his mind – his lucrative aspirin businesses.

Outside the United States, the most successful of these concerns was in Latin America where Bayer Aspirin enjoyed an effective monopoly. The terms of Sterling's deals with Leverkusen meant that 75 per cent of the profits from this business had to be returned to Germany, but the rest was pure jam. Weiss was determined to exploit the market as aggressively as he could. He gave Max Wojahn, the regional boss, permission to spend a small fortune on advertising and set about developing a new product especially for local consumers.

The result was a new brand, bearing the all-important Bayer trademark and made at Leverkusen. It was called Cafiaspirina (aspirin with a little energizing caffeine added) and was backed by the most ambitious advertising campaign that Latin America had ever seen – radio, newspapers, posters on billboards, buses, trams and trains – all followed up by an aggressive and highly mobile sales force that took Cafiaspirina out of the cities and into remote rural areas. They travelled in specially designed motor vans, each plastered with its own posters and hoardings, and sometimes took cinema projectors with them, stopping in village squares and forest clearings to set up a screen on which they would show newsreels, cartoons, comedies and, of course, short documentaries about their wonderful new drug.* Many of the farmers and peasant workers who saw these films (often a completely novel experience in itself) hadn't even heard of aspirin before and were astonished by the discovery that the little white pills that the Bayer reps pressed on them really *could* get rid of their aches and pains. Not surprisingly,

* A similar strategy was followed by Nicholas Proprietary that took the Aspro brands into some of the most inaccessible parts of South East Asia. Salesmen in countries like Thailand and Burma (still part of British India at the time) would travel up-country by riverboat and canoe, lugging film projectors with them. In parts of Indonesia, the first white people that some of the indigenous population ever met were selling Aspro. The cash returns were small for obvious reasons but the strategy built up a huge market for the future.

it wasn't long before these new customers were popping aspirin at every opportunity.

The returns were exceptional. Financially, of course, IG Farben did best out of the deal. Its 75 per cent share of the profits made the region one of the largest net contributors to the cartel's bottom line – $800,000 dollars in 1925, and over $1.25 million in 1929 (equivalent to several tens of millions today). Indeed, it was such a reliable source of income that it played a big part in helping the IG get through the worldwide depression that followed the Wall Street Crash. But Weiss was happy enough. Leverkusen had the added burden of manufacturing and shipping the drug, whereas all Sterling Products had to do was sell it. And selling was always his forte. He sat back and watched the profits pile up.

This avalanche of money may account for some of his curious naivety about what was happening in Germany. Throughout the 1930s much of his knowledge of events there came from the IG's head of pharmaceutical sales, Wilhelm Mann. As a loyal IG employee, Mann had colluded with Weiss in the arrangements to keep the complicated set of deals with Sterling secret because the Nazi Party had become ever more grasping in its search for donations and, for a time at least, the IG management wanted to keep the scale of the American profits out of the books. However, Mann was also a Nazi supporter and had consistently played down the more unpleasant aspects of the new regime, telling Weiss, for instance, that Germany's new rulers were pro-business and that there was no truth in any allegations that Jews were being persecuted. Because Weiss had learnt to trust Mann's judgement on commercial matters he seems to have accepted these other reassurances at face value too. If so, it was a dreadful error of judgement.

Alarm bells should have rung when the Nazi Party's overseas apparatus began to take an interest in the IG's Latin American interests. In the mid-1930s it dawned on someone in Berlin that the IG offices dotted around Brazil, Argentina, Paraguay and Peru could be useful adjuncts to its propaganda battle aimed at establishing Nazi spheres of influence among expatriate German communities, and because Cafiaspirina was essentially a German product (or so the Nazis thought at the time), they sought to extend their authority to its aspirin sales force as well. The IG had no alternative but to pass this pressure on to Sterling. Before long the swastika began to creep into the combine's Latin American advertising and even on to some Cafiaspirina posters too.

Matters got even worse in the spring of 1938 when the German regime's accountants finally learnt of the secret deals that Sterling and Leverkusen had struck in the 1920s. Because the two companies had successfully hidden the payments from these contracts, it looked to the Nazis as though Leverkusen had given away important patents and trademarks to a foreign concern without any recompense. In order to deflect criticism, the IG was forced to ask Sterling to make a show of handing over $100,000 in compensation. Weiss reluctantly agreed on the promise (never fulfilled) that his German partners would find a way to get it back to him in secret.

The more Sterling Products allied itself with IG Farben through deals like this, the more dangerous and problematic its position became. The international situation was worsening and another war in Europe now seemed likely. Weiss must have realized that if war came there was every chance that America would line up with its old allies against Germany. If that happened, it would only be a matter of time before the US authorities began investigating American firms that had traded with the enemy. Unless he was careful, he might even lose his much-prized aspirin business.

But now Weiss made one of the biggest mistakes of his life. Instead of cutting the ties with IG Farben while he still had time, he dug himself and Sterling deeper into trouble.

When war broke out in Europe in September 1939, the British began blockading German exports across the Atlantic once again and it quickly became clear that shipments of the hugely profitable Cafiaspirina from Leverkusen would be severely disrupted. The only way to get around this shortage was to start making it at Sterling's Bayer Aspirin factory at Rensselaer and export it from there instead. But Weiss knew from experience that the notoriously touchy IG would see this as an attempt to cut it out of one of its most lucrative markets. So he came up with a sweetener. He offered (through Wilhelm Mann) to take over the production and sale of *all* the IG's drugs sold in Latin America but to run these businesses secretly in trust for the Germans until the war was over. The IG board, which was becoming just as worried about the impact of the conflict, reluctantly agreed, on condition that absolutely binding contracts were drawn up to ensure that the status quo was re-established once normality returned. And so the deal went ahead, hidden like all their previous arrangements in an elaborate chain of front companies.

Unfortunately, there was one massive drawback. The new arrangements made Sterling, to all intents and purposes, a direct production subsidiary of a cartel that most of the world now saw as a key part of the Nazi regime.

There was a kind of inevitability about what followed. In mid-1940 the Luftwaffe began bombing London and American public opinion started to harden against Germany. First the Justice Department – prompted by J. Edgar Hoover's FBI – and then the US Senate announced that they were opening investigations into possible Nazi infiltration of American industry. And that's when Weiss found out that he hadn't hidden things as well as he thought. In May 1941, someone started leaking information to the press. The *New York Herald Tribune* published a lurid account of the Latin American pharmaceutical trade, which it claimed was being controlled by Berlin, and cited Cafiaspirina as its chief example. Other papers weighed in, publishing names of the companies that were suspected of 'helping to pay' for Hitler's war. Right at the top of each list was Sterling Products. Suddenly, everyone from the US Treasury Department to the Securities and Exchange Commission began to take an interest. One after another they all began investigations.

Weiss and Sterling twisted and turned under this scrutiny but it didn't take long before the government men began looking into the company files and uncovered the secret deals that Weiss had struck with the IG. He might actually have got away with it, if all the probing had revealed was his aspirin contracts – he had bought the Bayer name in good faith in 1918 and it would have been difficult to prove that he had done anything illegal. But the later agreements came to light as well and they couldn't be so easily explained away. He was soon being accused of everything from running an illegal cartel to collaboration with an alien power.

The business started to collapse under the pressure. Sterling's factories stopped shipping IG drugs south to Latin America, its assets were temporarily frozen by the Treasury Department, and even Cafiaspirina was sacrificed, Sterling having to promise to launch a new brand called Mejoral that would compete with the German-produced aspirin from which it had made so much money in the past. Weiss was devastated by this last move but his lawyers told him that the alternative might be criminal charges that could cost him his freedom. With a heavy heart he cabled Leverkusen to inform them that he had to dissolve every one of their agreements. IG Farben's furious reply, insisting on the total fulfilment of its contractual rights, went unanswered.

But if Weiss thought that he had done enough to appease his critics, he was mistaken. The Justice Department was determined to sanitize Sterling Products of any Nazi connections and demanded Weiss's head. In August 1941, he was banned from the company for life.*

He was distraught. In just a few months he had lost the business that he had spent forty years building up; even his prized aspirin contracts, which he had fought tooth and nail to nurture and protect, had been snatched out of his hands. Broken and humiliated, he retired to West Virginia where he died in a road accident a year later.

It is hard not to feel a sneaking sympathy for the carpet-bagging, entrepreneurial William E. Weiss. Duplicitous and grasping though he sometimes was, there's no evidence to suggest that he was ever actually a Nazi sympathizer or, indeed, that he really understood what was going on in Germany. His motives throughout all his dealings with Leverkusen had always been purely commercial; to secure the best terms of trade for Sterling Products and ultimately to safeguard its greatest asset – the rights to sell a hugely successful medicine in some of the most lucrative markets in the world. By a combination of huckster's luck and shrewd wheeling and dealing he'd managed to build a massive business on the back of this windfall, and had he severed his connections with IG Farben earlier he might have hung on to it. His biggest sin was that of naivety, of believing that business was just business and that the events of the wider world were of no concern to him.

But at least his fate had been in his own hands. Another of the aspirin set was denied even this privilege. He wasn't a salesman, or a business leader, but a scientist, one of those who had brought the drug into being so many years earlier.

The Hall of Honour of the Deutsche Museum was busy that day, but to the tall, good-looking man standing by some glass cabinets on the

* Alongside Arthur Diebold, his original partner and chief moneyman.

ground floor, the chattering groups of Hitler Youth and tourists might as well not have been there. He was transfixed by the two displays in front of him. Inside the first case lay a small mound of pure white crystals, accompanied by the inscription: 'Aspirin: inventors Dreser and Hoffman'. The next cabinet along was filled with another substance, a compound of vital importance to the manufacture of everything from photographic film to plastics. It too bore a caption, bearing both its chemical and trade names. 'Acetylcellulose – Cellit'. But there the inscription ended, with no mention of the inventor.

Eventually, the man turned and walked slowly towards the exit. As he went through the door he passed under a large sign that forbade entry to 'non-Aryans'. It was Munich, 1941, and Arthur Eichengrün had just made a startling discovery. He had been written out of history.

Perhaps, given the political atmosphere of the times and the other misfortunes which had befallen him as a result, Eichengrün shouldn't have been so surprised. But all scientists want their discoveries to be recognized by their peers, and to be so blatantly and publicly denied credit for his part in creating two of the most important chemical formulations of modern times was a cruel blow. The story of how that happened, and what it led to, is one of the sadder episodes in the long history of aspirin.

When Eichengrün joined Farbenfabriken Bayer in 1896, he was already an experienced chemist, with a doctorate from the prestigious Erlangen University and four years' work in the industry behind him. His chief task at Bayer was to build up the pharmaceutical laboratory and to find new drug compounds. In the early days of the department its collegiate atmosphere meant there was no formal hierarchy as such but it was widely acknowledged that he was its most senior scientist (a position formalized in 1901 when he was made head of pharmaceutical research.) Part of his role, therefore, was to give counsel and encouragement to his junior colleagues and to monitor progress on whichever projects they had on the go. The development of ASA, worked on by Felix Hoffman, had been one of those projects.

Whether Eichengrün had instructed Hoffman to take a look at new ways to acetylize salicylic acid, or whether Hoffman, motivated by a desire to ease the suffering of his rheumatic father, did so off his own bat and then later showed his results to Eichengrün, is a matter of some controversy (although, as will become clear, evidence points heavily towards the former). However, beyond a certain point, the question is academic because not even Eichengrün ever denied that that Hoffman had played *some* part in the discovery. What perhaps is more relevant is that once the compound was prepared, Eichengrün had had to fight hard against the intransigence of Heinrich Dreser, the stubborn head of Bayer's pharmacology section, to get it into production, even going behind Dreser's back to arrange secret clinical tests to demonstrate its efficacy. (It may be remembered that when Eichengrün had circulated reports of these tests Dreser had scribbled the words 'This is the usual Berlin boasting. The product has no value' on the memo.) Without Eichengrün taking this risk, Dreser's opposition would have been conclusive and aspirin would not have seen the light of day. Fortunately, Carl Duisberg heard of Eichengrün's trials and stepped in to give the new drug his backing. The rest, as they say, is history.

From then on, Eichengrün played no real further part in aspirin's development, except for his contribution to the round robin discussion that gave it its name, and left it to others in the company to sort out all the details of the drug's official clinical testing and its subsequent promotion, marketing and sale. He had plenty of other projects to be getting on with (during his time at Bayer he would be responsible for eighteen major pharmaceutical patents), and it was Dreser's job, not his, to write the key pharmacology paper that launched the product. Indeed, he seems not to have taken much notice of aspirin at all once it was out of his hands; partly because he wasn't entitled to any royalties from it, partly because it took a little while for it to become the huge commercial success it did. And by then he had moved on.

As for credit, well, it wasn't the done thing in the Bayer pharmaceutical lab to go around claiming the glory for something that had been a team effort, even if it had originally been your idea. Later, others outside the firm would recognize Eichengrün's key role – the *Daily Mail* did so in 1920, and the *Chemical Trades Journal* in 1929, and the biographical handbook of the chemistry trade, *Handbuch der deutschen Gesellschaft*, mentioned it in 1930 – and, of course, his family and colleagues in the industry knew the truth of it, which was probably all that mattered at the time. It certainly wouldn't have bothered Eichengrün that Felix Hoffman's name had been included on the US patent because someone had to put their name on it and the team who had worked on developing a drug often took it in turns to do so at such times. After all, the unsuccessful German patent application for ASA was in the name of Otto Bonhoeffer, a scientist who had nothing whatsoever to do with the discovery itself, but merely worked on a way of producing it economically.*

Eichengrün stayed at Bayer for another nine years, playing a leading role in many of their other inventions, contributing to the leading journals and generally earning a reputation as one of the top pharmaceutical chemists in the country. But by 1908 his interest had swung away from drugs towards a new field and he decided it was time to go it alone. For some years he had also been working on compounds connected to photographic film and plastics; he had invented a new chemical developer called Edinol and then a substance based on acetylic cellulose called Cellit, which was later used, among other things, to produce non-inflammable camera film for Eastman Kodak and Pathé.

Eichengrün saw an opportunity to exploit the commercial opportunities that this new branch of science presented and so he left Bayer and scraped together enough money to set up his own factory: the Cellon-Werke in the Tegel area of Berlin. It was a great success, both as a production plant and a hothouse for his ideas. Cellon (a more commercial variant of his previous discovery) was an early plastic – transparent, strong, and elastic – that could be used as windowpanes for cars and aeroplanes and later led to Cellophane. Other product modifications included plastic lacquers that could be painted on aeroplane wings, fire-resistant paint, artificial acetate silk and even the little plastic inlays which gentlemen put into their shirt collars to keep them stiff. He held all the patents to these inventions personally and was soon making as much money from licensing the technology to others as he was from making the goods himself.

As a result he began to live in some style, with a large house in the country, a flat in Berlin, a car, a yacht, paintings, fine furniture and books. But nobody seems to have resented his success. He had a strong sense of right and wrong and could be a demanding boss, but by all accounts he was also a good-humoured and generous man who enjoyed the company of his friends and family. He was regarded as handsome

* The British patent had been in the name of Henry Newton, the company's British nominee.

by the standards of the time, too – tall, confident, with a full moustache, firm handshake and a steady gaze – which probably accounted for much of his success with the opposite sex. Indeed, if Arthur Eichengrün had an Achilles heel it was his love of women. Even one of his grandsons remembers him as being something of a ladies' man.* He was married three times – the last time to Lutz Bartsch who had been a nanny to some of his six children.

And so the years went by, punctuated by new inventions, a glittering Berlin social life and Eichengrün's various love affairs. There were lows of course; his company hit hard times once or twice when the economy went into a spin, but he always seemed to pull things around without too much effort. If things ever got too pressurized or hectic he threw himself back into science, his abiding passion. He was never happier than when absorbed in some new project at the Cellon-Werke laboratory. All in all, it was a very pleasant existence.

As for the wider world, although Eichengrün must have been aware that Germany was slipping into political turmoil, like many others of his class he probably didn't take the National Socialists too seriously or believe they posed a threat to him personally. Fringe parties were always making trouble and common sense always prevailed in the end. But of course, this time common sense did not prevail. The Nazis came to power and from then on everything was different. Because Arthur Eichengrün was Jewish.

He had never been remotely interested in his religion. As a boy in Aachen he had done his family duties at the synagogue because his father, a cloth merchant, expected it, but at the first opportunity he had turned his back on spiritual matters. Like many other scientists of his day, he had no great belief in abstract notions of faith and no need of the institutions that held the Jewish community together. He had married outside it and brought his children up in the same way and, if he had any personal philosophy at all, it was a vaguely humanistic one that emphasized individuals' responsibility to follow their own way in life and make the best use of such opportunities and talents as they had. He sometimes even forgot he was Jewish at all.

Unfortunately, to the incoming Nazi authorities this kind of spiritual

* That grandson is Ernst Eichengrün, to whom I am indebted for the many personal details about his grandfather and his achievements that appear in these pages.

ambivalence was irrelevant. It mattered little if you were Jewish by inclination or not. In their eyes, once you were a Jew, you were always a Jew. Some Jews were clearly more threatening than others – practising Jews, obvious Jews, and they would be dealt with first. But anyone who thought they could simply walk away from their racial ancestry by pretending to live in the guise of good 'Aryans' would one day have to think again.

Little of this was immediately obvious to Eichengrün because his wealth and marriage to a gentile insulated him from what was happening to others. The anti-Jewish propaganda and speechifying were unpleasant and there were some awkward new laws that he now had to abide by, but with the Nazis keen to reassure German sceptics and anxious overseas governments about their credentials, it took a while for the true state of affairs to become clear. He was officially told that he would have to take on an 'Aryan' business partner at Cellon-Werke in order to maintain his many State contacts. But otherwise, if he kept his head down, he could continue to enjoy some of the other benefits of living in Germany – for now. So he got on with things as best he could. Indeed, there was a slightly surreal aspect to his life around this time because Hermann Goering acquired a plush apartment in the block where Eichengrün lived and he got used to seeing him in the lift. Goering, who clearly didn't have much of a clue who Eichengrün was, would regularly tip his hat to the scientist and his wife and on one bizarre occasion even gave some sweets to their niece.

But had he wanted to identify them, there were soon other signs of Eichengrün's diminishing status. In 1934, a recently retired IG Farben scientist called Albrecht Schmidt published a history of chemical engineering (*Die industrielle Chemie in ihrer Bedeutung im Weltbild und Erinnerungen an ihren Aufbau*). The book was an account of the world's great scientific innovations, with, as might be expected, much detail about German achievements in general and IG Farben's success in particular. On page 775 there was a footnote that told the story of the 'discovery' of aspirin by Bayer scientist Felix Hoffman. It was notable for two things. It was the first appearance of the legend that had young Hoffman being encouraged by his rheumatic father to produce a therapy free from the unpleasant side effects of salicylic acid. And it made no mention of Arthur Eichengrün's role in the affair.

The account begs a number of questions. The first is, where did

Schmidt get his information? Up until then various vague references to the paternity of aspirin had appeared in publications, the most concrete of which were those (listed above) that ascribed much of the credit to Arthur Eichengrün. Heinrich Dreser, on the other hand, had been named as the scientist responsible on at least two occasions – in 1920 and 1930 – by none other than Carl Duisberg himself (for reasons that are hard to explain unless he simply mixed things up or wanted to conceal the fact that Bayer's head of pharmacology had initially erred by refusing to accept aspirin's potential).

But until 1934 history had been silent about young Felix Hoffman's role in the affair. He had never published a word about aspirin himself, or indeed about heroin, his only other 'invention', and shortly after they were compounded he had left the laboratory to become head of pharmaceutical sales. He was still with the company in 1934, however, and it's possible that he was the one who told Schmidt this story, although it's hard to believe that he would have tried to claim sole credit in such a roundabout way when at any time in the preceding thirty-seven years he could have gone into print himself. Aspirin had become famous and as its inventor he would have had an eager and fascinated audience.*

Perhaps, then, Schmidt heard it from someone else in the company, or made it up himself? It's impossible to know now as he is long since dead, but he could have fashioned the story together from scraps of company gossip to make a nice neat tale. However, if this was the case, why would he have left Arthur Eichengrün out of it altogether – a man who was held in the highest regard by the scientific community, who had dozens of brilliant inventions to his name and who had been much better known as a Bayer scientist than Felix Hoffman?

Which leaves the possibility that other more sinister forces were at work. Did someone at IG Farben, or Schmidt himself, exclude Eichengrün on purpose? This was Germany, 1934, and the Nazis had recently come to power. Could someone within the company have

---

* Hoffman didn't publish anything about aspirin after these events either and remained silent on the subject until his death in 1945. In 1918, in an internal Bayer history, Eichengrün himself had mentioned in passing that Hoffman had prepared the compound but had left open the question of why Hoffman had done so – presumably because he thought there was no need to make this clear.

thought that it might not be the best policy to give some of the credit for one of the most remarkable pharmaceutical formulations in history to a Jew and leant on Schmidt to write a more acceptable version? Although this might seem unlikely, given that IG Farben's Carl Bosch was even then remonstrating with Hitler about the important role that Jews had played in German science, what later happened to Eichengrün means it cannot be totally dismissed.

At the time, Eichengrün made no attempt to refute the account and it is not even certain whether he saw it. If he did, it's possible that he dismissed it as a piece of research that was inaccurate or even just not worth bothering about. After all, those who mattered to him knew what he had achieved and he might have felt secure enough in his other successes not to care about this one. However, it's also possible that he could not contest it because of the position he was in. He had just had to hand over half his business to a Nazi supporter and, like every other German Jew in 1934, he may have felt it wiser to keep a low profile.

Whatever the case, the moment passed and from then on matters got much worse for Arthur Eichengrün. The Nazis' denunciation of the Jews grew louder by the day and business became more and more difficult as clients chose not to deal with a company that was even partly owned by a 'non-Ayran'. In 1938, he was forced to sell out Cellon-Werke entirely and moved his laboratory into his Berlin apartment – in the process handing over his many other investments to the government in return for worthless state bonds. Then one of the Nazi newspapers published a gossip piece asking why Hermann Goering was living in the same building as a Jew, and Eichengrün had to find somewhere else to live. Not long after came the Kristallnacht, when the Nazis smashed Jewish homes, stores and synagogues across Germany and Hitler announced that Jews would play no further part in the economic life of the country. By the time the war broke out and Eichengrün was regretting not having left the country while he had a chance, little of his once comfortable lifestyle was left.

Somehow he got by. The fact that his third wife, Lutz, was an 'Aryan' still counted for something, and he had friends who helped him out. Compared to many other Jews he was comparatively well off. But then in 1941 while on a visit to Munich, he managed to get into the Deutsche Museum and found out that not only was his part in aspirin's development uncredited, but he had also been denied any role in the invention of acetyl cellulose as well. Someone somewhere was determined to deny him his rightful place in scientific history. It brought home the stark reality of his situation in a way that nothing else could.

Meanwhile, the mass deportation of Jews was beginning to gather pace and the protective veneer of Eichengrün's mixed marriage was wearing thin. The authorities obviously had their eye on him and there were only so many times he could get through the endless security trawls unscathed. He had always to be on his guard against doing or saying anything that could get him into trouble.

Ironically, it was his wife who made the fateful error. In 1943, she had to write a letter to the authorities and the only paper she could find was a piece of Eichengrün's old business stationery. In accordance with the race laws he was supposed to have put the word 'Israel' in between his first and last names on the letterhead. But he had forgotten to do this. A clerk who noticed the omission reported him and he was arrested and sentenced to four months in prison. A few months after his release, he was re-arrested for the same offence (double jeopardy was a common occurrence for Jews in Nazi Germany) and this time he was deported to Theresienstadt concentration camp.

The regime at Theresienstadt was brutally tough, but in the perverse way that these things were measured it could have been worse. Situated on the Ohre river, near Prague, it comprised an old fortress with a walled village attached and was more like a ghetto than the purpose-built extermination camps found elsewhere. At one point the Nazis had even used it as a showcase to fool the Red Cross that conditions in the camps were tolerable, and at the beginning of the war, thousands of elderly Jews had been beguiled into *volunteering* to go there. Clearly the camp regime was corruptible too, because somehow, either through influential friends on the outside or by bribing a guard, Eichengrün was able to get a room to himself.

But he quickly found that behind this facade, the SS ran Theresienstadt with all the savagery they did elsewhere. It was freezing cold, shootings, beatings and torture were common, rations were at starvation levels and there were no medical facilities worth their name. This was particularly problematic for Eichengrün as he had recently been diagnosed a diabetic and was in serious need of the right care and drugs – a paradoxical position for someone who had made such a remarkable contribution to German pharmacology. Perhaps what kept him going was the knowledge that for many thousands of inmates the camp was just a staging post for even more dreadful destinations. Trains to Auschwitz left every few days and the guards were always able to find room in them for those who were too sick to work.

Eichengrün may have been sustained too by an overwhelming sense of injustice at the way his scientific achievements had been suppressed, because it was at Theresienstadt that he began his campaign to restore his reputation. He wrote a letter to his old employees at IG Farben setting out his role in the development of aspirin.

How this letter got out of the camp and found its way safely to Leverkusen in late 1944 is something of a mystery. The Soviet Army was advancing rapidly across Eastern Europe, the Allies were invading from the West and their bombers were battering away at Germany's infrastructure almost day and night. But get through it did and it remains in the Bayer archives to this day. It was too much for Eichengrün to expect any redress, of course, and in any case, once he'd written the letter he was more preoccupied with staying alive. That winter the Russians moved ever closer. On Heinrich Himmler's orders the Theresienstadt authorities took delivery of a consignment of Zyklon B canisters and began building a gas chamber. For the inmates it was a question of which would come first. Fortunately the Red Army beat the Nazis to it by a couple of weeks, but it was too late for the 34,000 prisoners who had perished at the camp in the preceding two years, or the 83,000 who had passed through it in on their way to the death camps in Poland.

When Eichengrün finally managed to get back to Berlin in late 1945 (he and the other inmates had had to walk forty miles to Prague and then wait weeks for transport to appear), the city was almost unrecognizable. His wife, Lutz, was still alive, but the SS had destroyed his apartment building during the last weeks of fighting and all their possessions had gone. It might have been supposed, given his deteriorating health, the appalling privations that he had suffered and the shortages that he now faced, that he would have wanted to put the past behind him. But Arthur Eichengrün's scientific reputation meant everything to him. He felt he had been denied the recognition he deserved because of his Jewish ancestry and now, having miraculously survived the Holocaust, he was determined to put that right. In any case there was little else to do – he quickly found it was impossible to reconstruct his business, and even private laboratory experiments were out of the question because of the shortage of materials. So he and Lutz moved south to the more temperate climes of Bavaria and there he began to write his valedictory paper.

It was published in 1949 in a small journal called *Pharmazie*, under a title that celebrated an important anniversary: 'Fifty Years of Aspirin'. He told the same story he had set out in his letter to IG Farben – a concise, straightforward but convincing account of how aspirin came into being, and of the respective roles played by Felix Hoffman, Heinrich Dreser and himself. It was clearly the work of a man who was in command of his facts, someone who had had the time to recall the exact sequence of events as they had been played out all those years earlier. It was all there: how he had instructed Hoffman to synthesize acetylsalicylic acid and how Hoffman had done so without knowing the reason for the work; how, faced with Dreser's refusal to conduct clinical studies, he had tried the compound on himself and then arranged secretly for it to be tested by Dr Felix Goldmann and other physicians in Berlin. Dreser had dismissed these successful results as loud-mouth boasting, but then Carl Duisberg had stepped in and ensured the results were independently checked. Aspirin was found to be satisfactory.

The article appeared in December 1949, and a fortnight later, aged eighty-two, Arthur Eichengrün died. He must have gone to his grave believing that he had set the record straight and that the world would once more acknowledge the key role that he had played in developing its most remarkable drug.

Tragically, his article would be ignored for another fifty years. The new Bayer that emerged from the wreck of IG Farben after the Second World War would continue to maintain that Felix Hoffman was the drug's only inventor and that Heinrich Dreser had brought it into production. This was still its position in 1999, the year of aspirin's centenary, when no mention was made of Arthur Eichengrün in all the celebrations and attendant publicity. It wasn't until Dr Walter Sneader, a tenacious medical historian from Scotland's University of Strathclyde, began asking questions and digging into the archives in preparation for a paper he was giving, that something like the true story began to re-emerge. Even to this day, the company will only grudgingly concede the possibility that Eichengrün may have played a part in aspirin's history, and a very small part at that, which, given the weight of evidence in his favour, is hard to understand.

By the mid 1940s IG Farben's association with the Nazis had become so complete that it was just another moving part in the Third Reich's machinery. Hitler's total war demanded total commitment from the German people and like so many other individuals and organizations the IG gave itself up to him. There were some instances of redemption – amid all the munitions, gases, synthetic petrol and rubber and the other materials that the IG produced to sustain the war, it also continued to develop and manufacture important drugs that saved lives and healed and comforted the sick. And there must have been a number of IG employees – workers, scientists, salesmen, even executives – who were as sickened as every other silent German sceptic by the madness of the adventure that the Führer had led them into, but were powerless to do anything other than see it through to the bitter end.

Corporately, however, the combine lost its soul. What had started out as a bold, aggressive and expansionist enterprise (no more illegitimate in its aims and ambitions than say, General Motors or British Petroleum) became a tool of murderers and plunderers. Many of its employees didn't just toe the Nazi Party line, they actively embraced it, and bent all their efforts to doing its work and advancing its cause. This was the largest and most powerful industrial combine in Germany, if not the world, and its active participation in the execution of Hitler's plans led to acts of manifest evil.

To take just three of the many products of this collaboration: the cartel provided several million marks for the construction of the Auschwitz extermination camp, with all its grotesquely inhuman consequences. It also built and ran an adjacent chemical plant, IG Monowitz, where tens of thousands of slave labourers died because of the appalling working conditions, ill treatment and malnutrition. And it financed and participated in medical experiments conducted by Nazi doctors and scientists on concentration camp inmates, resulting in the torture and death of thousands of victims.

One tragic story will be enough to serve as an illustration of IG Farben's moral collapse during these years. It concerns a pair of ten-year-old identical twins, Eva and Miriam Mozes, from the small village of Portz in Romania, who arrived in a packed cattle car at the Auschwitz railhead in March 1944. As the SS guards screamed at them to get out of the train, they were separated from their father and two older sisters. Their mother clung to them for a few desperate minutes until a soldier asked if her daughters were twins. When she said they were, the girls were hustled away. They never saw their mother again.

After a further horrific introduction to the realities of Auschwitz (on her first visit to the barracks latrine, Eva found corpses lying abandoned on the floor), the sisters discovered why they had been singled out. It was because Nazi scientists, led by the infamous Dr Joseph Mengele of the SS, had a liking for experimenting with identical twins. They were the ideal human guinea pigs – one could act as a natural control for the other. Mengele's personal obsession was with the Nazi ideology of racial purity and he believed that by working on twins he could unlock the secrets of reproduction and thereby help the 'Aryan' race repopulate the world. But he and his colleagues had other tasks at Auschwitz as well, including testing various important prototype drugs on behalf of IG Farben's Bayer pharmaceutical division.

Over the following months the Mozes girls were among 1,500 sets of twins at Auschwitz on whom such experiments were carried out – they were variously castrated, blinded, beheaded and deliberately infected with disease. Over the course of several months Eva and Miriam were given hundreds of injections of unknown substances, probably including an experimental typhus treatment (batch number BE 1034, found later in the Auschwitz laboratories), which was produced by the IG's Bayer pharmaceutical division. After one such set of injections Eva developed a raging fever and her limbs ballooned to several times their normal size, although it was hard to be sure whether this was as a result of the drug or because SS technicians had tied her down with rubber hoses to keep her still. Along the way the twins had several encounters with Dr Mengele. At one point he stood over Eva's bed, laughing as he read her fever chart. 'Too bad, she's so young,' he said to his colleagues. 'She only has only two weeks to live.'

Of course, despite the seemingly inexhaustible supply of human guinea pigs, Dr Mengele's experiments didn't come cheap, and IG Farben seems to have been willing to fund them. As Wilhelm Mann, the executive who had overseen the cartel's aspirin deals with Sterling Products, said in a letter to an SS contact at Auschwitz, 'I have enclosed the first cheque. Dr Mengele's experiments should, as we both agreed, be pursued. Heil Hitler.' A few months later, Mann was put in an ideal position to find out personally where the cartel's money was going – he became the chief supervisor at the nearby IG Monowitz slave-labour factory.*

Other IG staff had more direct involvement, including Dr Helmuth Vetter, a longtime company employee and an SS doctor at the camp. In 1943 he conducted research on 200 female prisoners, injecting the women's lungs with streptococcus bacilli and causing them to die from pulmonary oedema. His paper on these experiments was later incorporated into a presentation sponsored by the Military Medical Academy on the effectiveness of the new drugs being produced by the IG's Bayer pharmaceutical division. Vetter seems to have enjoyed his job. 'I have thrown myself into my work wholeheartedly,' he wrote to colleagues at Leverkusen, 'especially as I have the opportunity to test our new preparations. I feel like I am in paradise.' He was subsequently executed as a war criminal.

Eva and Miriam Mozes were among the luckier victims of the Auschwitz experiments, being among the 200 out of 1,500 sets of twins who survived to tell the world about their experiences. Luck, of course, is a relative term. As an adult Eva suffered from miscarriages and tuberculosis and Miriam's kidneys never developed properly – she later died of cancer. After spending some time in Israel, Eva eventually moved to Terre Haute, Indiana, in the United States. She is still alive today and runs a small museum devoted to the victims of the Holocaust. In February 1999, she was one of hundreds of survivors who joined in a class action suit against German pharmaceutical companies, including Bayer AG, for providing toxic chemicals for the concentration camp experiments and using information from those trials to make and market commercial drugs. She said recently:

> Emotionally I have forgiven the Nazis, but forgiveness does not absolve any perpetrator from taking responsibility for their actions. I am free. I refuse to forever be a prisoner of Auschwitz. I am free. But they are not and they never will be until they accept their responsibility. They are different people today.

---

* He was also on the board of the IG's Degesch subsidiary that produced Zyklon B.

> I know that the ones who ran Bayer fifty years ago are all dead now. But the company today should have the courage and decency to admit their past.

She added that she has always tried to avoid taking any Bayer drugs, even humble Bayer Aspirin.

As for the many other IG Farben employees who became involved in Nazi atrocities, only twenty-three of them (all senior executives) were eventually tried for war crimes at Nuremberg in 1948. Eleven of them were acquitted. Among those who walked free was Wilhelm Mann, who had managed to convince the judges that he had been acting under duress. Among the convicted were Carl Krauch, Chairman of the IG board, and Fritz ter Meer, a senior production manager who was sentenced to seven years' imprisonment for his part in building the Monowitz factory. During the trial ter Meer had spoken about the human experiments, saying, 'Concentration camp prisoners were not subjected to exceptional suffering, because they would all have been killed anyway.'

After Nuremberg, the IG Farben cartel was broken up by the Allied powers and three new companies arose from its ashes: Hoechst, BASF and Bayer. Bayer reverted to producing pharmaceuticals, pretty much as it had done when Carl Duisberg was in charge. Aspirin, the drug that had started it all, continued to be one of its most profitable and successful products. In 1956, not long after his release from prison, Fritz ter Meer became the company's new Chairman.

BOOK THREE

10

SOLUBLE SOLUTIONS AND COSTLY COMPETITION

IT WAS A dismal day to arrive. As the lorry laden with his furniture inched carefully through the streets, the mist parted momentarily to reveal bomb-shattered buildings and large piles of rubble. The few people out and about were heavily muffled up against the cold and from somewhere in the direction of the river the mournful groan of foghorns could just be heard. On that damp, dispiriting afternoon in November 1945, George Colman Green gazed out at his new home and wondered if he was doing the right thing.

The city of Hull had taken a pounding during the war. Within easy flying reach of Germany and with some of the busiest docks in the north-east of England, it was a natural target for the Luftwaffe and they had returned to it time after time. The bombers were now long gone, of course, but much of the debris remained, bulldozed to one side until the authorities had the time and money to start rebuilding. However, in other ways Hull was getting back on its feet again. It was home to one of the biggest deep-sea fishing fleets in Europe – at last resuming the arduous journeys to Arctic waters that the hostilities had interrupted – and to numerous other businesses, large and small, which valued the hard-working Yorkshire populace and the ease of communications offered by the Humber river. One of those companies was Reckitt and Colman, venerable manufacturers of Brasso, Dettol, Robin Starch and sundry other items indispensable to the British housewife. With the war over it was about to embark on a new venture and George Colman Green was coming to take charge of it. For the second time in twenty years, Hull was about to make its mark on the aspirin trade.

Green's arrival was the culmination of a long train of events that had begun seven years earlier in the more refined surroundings of Guy's Hospital, London. At the beginning of 1938 one of its leading physicians, Arthur Douthwaite, had been experimenting with a device called a gastroscope, a long thin rubber tube fitted with tiny lights and mirrors that could be poked down a person's oesophagus. Invented in 1932, it allowed doctors to examine the lining and the wall of a patient's stomach without having to resort to surgery. The gastroscope was a fiddly piece of equipment that took much practice to operate but during one of Douthwaite's many trials he came across something very unusual. The patient he was examining had recently taken an aspirin and Douthwaite found a fragment of it, still gripped in the folds of the stomach lining. What's more, the tissue around it was bright red and very inflamed. The implication was obvious – the aspirin was causing the irritation. After further tests confirmed his conclusions he wrote them up in the *British Medical Journal* and it wasn't long before other doctors began drawing on his observations to solve various unexplained medical problems. The following year, another eminent physician, Sir Arthur Hurst, wrote to the *Lancet* with an account of a patient who had been vomiting blood at home, but who stopped doing so when he came into hospital. Puzzled by this, Hurst had gone back over the man's history and discovered that he had taken aspirin at home, but had ceased to once he was admitted to hospital. Hurst remembered Douthwaite's paper and had a brainwave. He gave the patient an aspirin and then used a gastroscope to examine the effects. The cause of the gastric bleeding was revealed.

For many years doctors had suspected that salicylates could irritate the stomach, although they didn't know exactly how. Indeed, aspirin had been invented in response to this very side effect and until Douthwaite's observations it was thought to have successfully eliminated the problem. By and large, of course, it had. The vast majority of people who took aspirin suffered no problems at all unless they took excessively large doses. But it was now clear that a few people at least were still extremely sensitive to it and even, in the very worst cases, liable to haemorrhage. Although these sufferers seemed to be very rare, with hundreds of millions of people using the drug on a regular basis, it was a matter that couldn't be ignored.

Part of the problem seemed to be that some brands of aspirin dissolved faster than others, leaving irritating fragments attached to the stomach wall. During subsequent gastroscope experiments Douthwaite had tried giving some patients calcium aspirin instead, as this was believed to dissolve more quickly. Sure enough, in those cases the inflammation either did not appear or was much less severe. But calcium aspirin had one major drawback; it decomposed with storage. The tablets crumbled or became soggy very quickly, liberating acetic and salicylic acid separately, which defeated the whole point of aspirin in the first place. Aspirin manufacturers on both sides of the Atlantic had been trying for years to find a way around this problem because having a product that dissolved more quickly, and could therefore be said to be 'faster-acting', would be a good selling point. But none of them had been able to make aspirin that didn't decompose.

Just before the war, the boss of one small British aspirin manufacturer took this conundrum to Reckitt and Colman, a company that had won a reputation as a chemical innovator because of its success in producing skin-friendly antiseptics. The discussion quickly fell apart over money, but in the meantime it whetted the commercial appetites of Reckitt and Colman's managers who asked their chief chemist, a Mr Stevens, to investigate. Unfortunately, the war broke out a few weeks after Stevens began work and like many other employees he decided his duty lay with the armed forces instead. The project was shelved.

Then one night in July 1941, the Luftwaffe launched a fierce attack on Hull. It isn't clear if they were targeting the docks and a few bombs went astray, or if they were aiming directly at industrial sites in the city. In any event, Reckitt and Colman's premises in Damson Lane received several direct hits. No one was hurt, but the bombing destroyed the company's research laboratories with all their records, equipment and instruments. One of those whose work was lost was an expert in carbon black stove polishes called Harold Scruton, who had been trying to find new ways to reformulate the products with the few chemical resources the war effort allowed for non-essential manufacture. Now, with all his efforts lying in ruins and less than a year to go before he retired, he asked to be reassigned. His bosses didn't have much to offer him, but to keep him occupied they suggested he pick up where Stevens had left off. Why didn't he see if he could come up with a small aspirin tablet that dissolved easily in water, would be reasonably palatable and, above all, stable enough to last?

Scruton was well aware that this was easier said than done. He was a trained industrial chemist but for the previous forty years he had been working almost exclusively on household products. Neither he nor the company had any real experience of working with medicines, and with the laboratories lying in ruins, it was clear he would have only the most basic facilities to work with. However, he told himself, it was wartime and everyone had to make sacrifices. If nothing else, it could be an amusing challenge to while away the last few months before his retirement.

So he salvaged some bits and pieces of equipment, moved with the other scientific staff into an old laundry – one of the few buildings still standing – and doggedly went to work. Like George Nicholas many years before, he had first to re-acquaint himself with the basic chemistry, although as the formula for ordinary aspirin could now be found in any chemistry textbook that didn't take long. A trickier problem was finding out why previous soluble aspirin prototypes had proved so degradable and what could be done to prevent this happening.

It was, he soon realized, all a question of timing. The initial problem was combining the right substance or base with aspirin to make it into a soluble salt. Crucially, this must also avoid precipitating a reaction that would break the drug down into its constituent elements (acetic and salicylic acid) before they had been absorbed together into the bloodstream. He knew that salicylic acid on its own caused gastric irritation and was nowhere near as effective as when joined with acetic acid: it was vital that they passed through the stomach wall together. In other words, he had to find a base that would not react with aspirin until the last possible moment. Then, of course, it must work very fast indeed, and be completely soluble to boot. It was a tricky chemical problem made even more complex because of the way aspirin reacted to water. Water – even moisture in the air – could be anathema to basic aspirin, breaking down into its acetic and salicylic acids faster than anything else. Scruton's chosen base had to temper this process too.

Up until that point most attempts at creating soluble aspirin had used calcium salt as a base because it dissolved very quickly, but unfortunately it degraded aspirin quickly as well. In the United States, Alka-Seltzer put aspirin together with bicarbonate of soda, but Scruton rejected this combination because of the apparent problems of keeping the bicarbonate stable in storage (the very earliest Alka-Seltzer tablets had sometimes exploded in their packets). In the end, after much trial and error with various alternatives, he decided that the only base that could possibly work was calcium carbonate or plain old precipitate chalk.

Getting the right amounts of chalk and aspirin together into one tablet was a nightmare, however. Too much chalk and it didn't dissolve fast enough, too little and the degradation process would kick in before the drug could be absorbed into the bloodstream. Scruton ran endless experiments using various grades and shapes of chalk particles (even adding some old laundry starch that he found around the factory), before mixing them with aspirin powder into a tin that he then rolled backwards and forwards along his workbench. Unfortunately, this process combined to produce a very repetitious noise – the swish-swish of the rolling powders punctuated by the sound of his measured tread as he walked up and down – which infuriated the other chemists working in the same cramped room. It was even worse when Scruton began hammering out sample pills on an ancient tablet press he had dug up from somewhere. But he had the bit between his teeth and ignored the complaints. Day followed day, week followed week, month followed month. The date of his retirement came and went, and still he worked on, mixing his powders, thumping out his tablets and then testing them to see their solubility. It was a mechanical kind of science with little immediate reward, but it was the only way he knew how to solve the problem of finding the right combination of ingredients to do the trick. And then, one day in February 1944, he decided he had cracked it.

On 24 February, he passed the results on to the Reckitt and Colman board for consideration and congratulations, although he may have been a little over-optimistic in his claims. He told them he had come up with a formula for a tablet that was palatable, stable in storage and highly soluble. It was ready to go for immediate clinical and consumer trials. He did concede that there might be one or two production problems still to resolve, but he didn't think these were insurmountable and in any case, the sooner it was tested the better. At a meeting of the company's research committee a month later, someone dryly pointed out that 'marketing will not be possible until after the war because of the lack of supplies of essential ingredients', but none the less, everyone else seems to have accepted that the project was ready to go. At long last, Harold Scruton felt able to retire.

Once he had gone, a small batch of tablets was sent out to the medical community for testing, but nothing much else happened for over a year. Try as it might, Reckitt and Colman couldn't find a suitably qualified scientist to take on the venture. Many of its staff were still away in the forces or otherwise engaged in vital war work and there was no one immediately available who could both manage the project through to production *and* understand the science involved. For fourteen months things remained on hold. And then someone remembered George Colman Green.

Before the war Colman Green had worked for one of the company's small subsidiaries, the Suffolk Chemical Company Ltd in Ipswich, but on the outbreak of hostilities he had been seconded to the Ministry of Supply. He spent most of the next five years in charge of an operation that produced all Britain's reserves of morphine for the armed forces, and then became part of a government team that followed Allied troops across Europe in order to uncover German scientific and industrial secrets. When the war finished he was at something of a loss, not sure whether he wanted to return to a humdrum life in Ipswich or even if his old job was available, and so when Reckitt asked him to go to Hull to take charge of what appeared to be a viable and interesting new project he jumped at the chance.

If his first impression of the bomb-damaged city hadn't been good, his initial appraisal of Scruton's work was even worse. He inherited his predecessor's bench in the cramped old laundry and large piles of hand-written notes but almost immediately frustration set in because he couldn't make head nor tail of the specifications they contained. He tried repeating some of Scruton's experiments to get a 'feel' of what he'd been up to, but despite several attempts he was unable to re-create a reliable tablet. He began to fret:

> It become obvious to me at an early stage that so unorthodox was his formula and his specification for some of the raw materials, it seemed hardly likely that his mixture could be [produced] economically on a commercial scale, if at all. But every attempt to 'bend' the formulation in the direction of orthodoxy led to slowness in dissolving, loss of palatability, loss of stability and general destruction of the characteristics that had been aimed at.

At best, accepting Scruton's formula meant he could produce only a few good tablets and a frightening amount of waste. At worst, the whole project was ill conceived from start to finish. He was even tempted to tell the Reckitt and Colman board that he was never going to make it work. But then some reports came in of the first clinical trials on the leftover quantities of Scruton's version of the drug (carried out at a London teaching hospital), closely followed by the result of trials at Birmingham Children's Hospital that Colman Green had commissioned himself using the few pills he'd been able to make to Scruton's formula. Both were uniformly positive. The drug was much easier to administer than ordinary aspirin, particularly to rheumatic children who needed high doses. Birmingham's physicians even asked for an immediate delivery of 20,000 pills.

The order caused tremendous excitement within the company and Colman Green found himself in an awkward position. He had to inform his bosses that there was no question of being able to make such large quantities of the pills in the foreseeable future. He had no plant, no equipment, and he still hadn't been able to get his head around Scruton's formula.

> The news . . . was accepted with incredulity in High Places. They had formed a false impression of the prospects because of failure to grasp the full import of the reservations that Mr Scruton had clearly made in his final report. There was final acceptance of the position, of course, as there had to be, but I felt that my reputation had not in the least been enhanced by this situation.

He was faced with a dreadful dilemma. Harold Scruton's recipe was clearly *medically* acceptable, but he just couldn't see how it could ever be made into actual tablets using conventional mass-production processes. The test pills were all made up by hand, with the chalk and aspirin powders mixed together before being poured into a die for pressing. However, this frequently went wrong – getting the density exactly right was a tricky business – and was incredibly wasteful of raw materials. For every tablet he successfully produced, dozens more had to be rejected. If the procedure were ever to work mechanically, let alone comply with the stringent requirements of the *British Pharmacopoeia*, every pill would have to contain exactly the same ingredients. But unfortunately the powders didn't flow freely when mixed, being more like flour than sugar in consistency. Getting exact measures every time was nigh on impossible. The only way he could see to solve the problem was to put the mixture through a process known as 'slugging' in which the powder was highly compressed into a very large cylindrical tablet. This would then have to be 'kibbled', or broken up into large particles, and sieved. Only then could it be poured into a hopper for dispensing to a tablet machine.

As anyone in the drug industry could have told him, 'slugging' and 'kibbling' required highly specialized machinery, and such things were simply not available in Hull in 1945 – or anywhere else in the country for that matter. When he made enquiries of a few manufacturers he was informed that it would take five years at least from placing the order to receiving the equipment.

Colman Green spent long hours at his workbench trying to find a solution, and in the end realized he would have to do what every other British scientist, engineer and inventor has had to do when faced with inadequate equipment. He would have to improvise. His first attempt was with some old presses (once used for making bath salts) that had survived the bombing, but they couldn't be effectively adapted. Then he went along to the company's engineering department and asked for help. Someone there knew of some single-punch presses that had previously done service in a government munitions factory. With the addition of a rickety belt drive, much tinkering and some hazard to life and limb, Colman Green managed to get this Heath Robinson contraption to work. But it was only the start of his problems. The powders would still have to be mixed and pressed and dried, each stage of the process requiring yet more equipment that wasn't available and yet more make and mend with barely suitable alternatives. He did his best, though, and soon strange pieces of machinery started arriving at the laboratory door, scrounged from here, there and everywhere; a batch mixer based on a design he'd once seen in Germany, some old sterilizing ovens from an army hospital, some hammer mills adapted from lab equipment.

The idea was to assemble all these bits and pieces into a pilot plant that would sit in the corner of the factory – a sort of test production rig that would allow him to see if he had the basic principles right. But putting it all together was one thing, getting it to work was fraught with difficulties. Either it would overheat and have to be switched off to stop the motors disintegrating or it would generate too much humidity, which meant the powders became a sort of glutinous mess. Eventually Colman Green and his long-suffering assistant, Fred

Dook, managed to keep it going but they were always on tenterhooks for when the next breakdown would occur.

In the meantime, the clamour for the product was growing. Experimental pills were still being sent out for testing in hospitals around the country and the response continued to be excellent. So certain were the company's managers becoming about its likely commercial success that in July 1946 they even put an order in for a full production plant. The equipment couldn't be delivered for another four years, but when up and running it would have a production capacity of 1.8 million tablets a week. Colman Green was aghast at this hubris. To make such a huge financial commitment when he hadn't even managed to get his pilot plant running reliably was a fantastic gamble. He couldn't even guarantee that he could make enough pills for 200 test consumers to try out the following spring. But he took a deep breath and went back to his machinery.

There were many other obstacles to overcome. The first was what to call the drug. That was easy enough: it would be called Disprin, reflecting its active ingredient and its soluble nature. Trickier was the question of its appearance. Initially, the plan was to make a biconvex tablet, one half chalk, one half aspirin, but when Colman Green tried to make one he found that the parts separated in the bottle. In the end he came up with a distinctive-looking parallel-faced, bevel-edged pill that everyone seemed happy with. The next nightmare concerned the container. Huge quantities of glass bottles were ordered before someone realized that coal, which was vital to glass manufacture, was still in short supply. The Bevan Boys, who had sustained the mining industry during the war, were now returning to their old jobs and there simply weren't enough replacements to work down the pit. It was several weeks before the bottle suppliers could confirm that they would be able to produce enough. Then a major advertising campaign had to be planned, and the price of a bottle had to be decided. As each of these matters was settled the launch of Disprin came one step nearer, and Colman Green's anxiety about his rattling equipment increased. Every now and then he would nip out at lunchtime for a quiet game of snooker in a nearby pub, but it was difficult to relax: would he be able to get it working reliably in time?

The fateful day was set for 20 November 1948, when Disprin would be introduced at the London Medical Exhibition. If it was successful there, a full public launch would follow and a major advertising campaign would kick in. Every doctor in the country would be sent a complimentary bottle with an explanatory leaflet in advance. To make enough tablets for them and to send out sufficient quantities to chemists to meet anticipated demand meant running the equipment at full blast, regardless of the risks. George Colman Green was alarmed, but by now he had begun to accept the inevitable.

> So many technical problems had still to be resolved and no full production run on the pilot-plant had up to that time been possible, that I felt that the commercial people were running away ahead of the realities of life. I made cautionary noises but I felt unable actually to oppose the Committee's recommendation. The gamble – which it was, technically – just had to be made and just had to have a favourable outcome.

Somehow he nursed the ailing equipment through the following weeks. On the opening day of the exhibition he was there on the stand with the company's sales team. It was decorated with a distinctive blue logo, split down the middle by a white sword; the same design that lay on the front of each bottle. The response was everything that the company wished for, and everything that Colman Green privately dreaded. Disprin was a huge hit.

The months following the exhibition were a sore trial for those running the unit back in Hull. The pilot-plant had been set up to solve manufacturing problems but was now having to play the role of a production unit on a scale which had never been anticipated. Demand from doctors, and then (with the launch of direct consumer sales) from the public, exceeded everyone's wildest expectations. The factory began to run double shifts, with those tending the plant having to leave the room at regular intervals to avoid building up humidity that would damage the drug. But somehow Colman Green kept the apparatus going until the long-awaited replacement machinery began to arrive. 'I have often wondered how many members of the board realized what a close call it was,' he said later.

He had achieved something of a miracle. In just over three years, amidst all the restrictions and difficulties of a country still recovering from war, he had taken some rough and ready chemical formulations and a few hand-pressed pills and produced what would soon be one of the best-selling analgesics in the country. He built a mass-production plant out of a few bits and pieces of redundant equipment in a half bombed-out factory owned by a company with no practical experience of making medicines at the time, and then by sheer willpower managed to keep it running in the face of the most relentless commercial and technical pressures. The scientific breakthrough behind soluble aspirin had undeniably been Harold Scruton's but his successor had been the one to turn it into reality. Without George Colman Green, Disprin might never have existed.

The product went on to be hugely successful in Britain and overseas. For the next few years it was the only proper soluble aspirin brand on the European market and it dominated much of the opposition. Its 'Take an Aspirin – I mean a Disprin' slogan achieved a prominence that is hard to imagine today. Even when Scruton's formula found its way into the *British Pharmacopoeia* in 1952 and others began to copy it, it retained much of its market share.

But, although it wasn't apparent at the time, the development of Disprin also came just as aspirin's star was beginning to wane. For the preceding half-century it had been in a class of its own – the only really effective over-the-counter painkiller on the market. It had survived (even thrived because of) wars, blockades, epidemics, personal ambition, political intrigue and commercial chicanery. It had made its many producers vast sums of money and by continually stoking demand through aggressive advertising they had done their best to keep things that way. Indeed, in the next two decades the rivalry between these producers would reach its absurd zenith as they fought desperate battles to convince consumers, erroneously, that there were fundamental differences between their various brands.

However, aspirin's supremacy couldn't go unchallenged for ever. Even as its makers were pulling on their gloves for the next round in the marketing ring, other pharmaceutical companies were looking for alternative analgesics. And when they found them, the competition that had sustained aspirin for so long would soon be threatening to kill it.

It is ironic that the first company to start eating away at the massive aspirin market had once been part of the enterprise that created it. Bayer Ltd had been the British outpost of Leverkusen but as a consequence of

Carl Duisberg's controversial deals with William Weiss, ownership had been split 50/50 between IG Farben and Sterling Products. In 1949, when the Allies were dismembering the huge German cartel, Sterling Products bought out the IG half of the business that had been confiscated by Britain's Board of Trade.

The move was part of a concerted effort by Sterling's new managers to re-establish its international position in the wake of the embarrassing revelations about its pre-war connections with Nazi Germany. In the process, Sterling had also radically reorganized the way its constituent business conducted their affairs, and in Bayer Ltd's case this involved distributing most of its product line among other newer subsidiaries. All that was left were a few ethical drugs and its aspirin business.

Unfortunately, that aspirin business was failing. Before, during and after the war, Aspro, Disprin and others had eaten into Bayer Ltd's market share to such an extent that by 1950 its annual sales amounted to only a few hundred thousand pounds. If the company was to survive, either a new marketing approach or a new product was needed. Bayer Ltd chose to go for a new product.

Although aspirin had completely dominated the world analgesic market for over half a century, there were in theory many other pain-killing substances around. Some of these shared similar chemical origins with the little white pill, going all the way back to the first aniline derivatives identified in Germany in the latter half of the nineteenth century. They shared another attribute as well – complicated chemical names that were shortened for simplification or easier marketing. One of the oldest of them was antipyrine, discovered by Ludwig Knorr in 1883. Then came Kahn and Hepp's acetylation of aniline, shortened to acetanilide and sold as Antifebrine. Another was Bayer's first success, made from waste para-nitrophenol, generically christened acetophenetidine in 1888 and subsequently sold as Phenacetin.

But although profitable in their time, these alternative synthetics had fallen out of favour over the years, either because they were thought to have harmful side effects or because mighty aspirin had swept all before it. If they were used at all it tended to be as additives to aspirin rather than alternatives to it, and even then they weren't therapeutically significant, just extra bits of window-dressing to help products stand out in a crowded market. Anacin, for instance, one of America's big new aspirin brands (sold as Anadin in Britain), contained additional acetanilide and caffeine when first formulated, although it was later reduced to just caffeine and aspirin. Another, Excedrin, contained aspirin, acetophenetidin, caffeine and traces of a substance called salicylamide. In both of these products, aspirin did most of the work.

However, Bayer Ltd's search for a new analgesic took it back beyond the origins of any of these substances to a half-forgotten aniline derivative with the convoluted chemical name of N-acetyl-para-aminophenol. This had been first synthesized in 1878 and was briefly thought to have as least as much therapeutic potential as acetophenetidine (its name was even shortened to the more manageable acetaminophen – normally a sure sign that someone hoped to make money out of it). But after tests apparently showed that it might have unpleasant side effects, none of the drug companies wanted to take it up and it was confined to the laboratory shelves as another of the many might-have-been substances that never quite made it into full production.

It reappeared in 1946 when Yale University scientists did some research into acetanilide. To their surprise, acetaminophen showed up as one of the substances that acetanilide metabolized into while in the body and actually seemed to be responsible for the latter drug's analgesic properties. Two years later, New York University researchers found that acetophenetidine also morphed into acetaminophen in the body. They too noticed its analgesic potential and, perhaps more importantly, couldn't find any signs of the supposed side effects that had blighted its earlier development.

Bayer Ltd's managing director, Laurie Spalton, was intrigued by these reports and asked his own scientists to investigate. They quickly came to the same conclusion as the Americans and told him that the research might prove the basis for a new product. From that point to clinical trials and from there to the creation of a new analgesic brand was a relatively straightforward process. The drug, given the name Panadol, was launched in Britain amidst much fanfare in 1956.

Paradoxically, Panadol's early success had much to do with Disprin, which by then was on its way up the British analgesic bestseller lists. Reckitt and Colman had based part of its successful Disprin advertising campaign on the pitch that soluble aspirin was not as irritating to the stomach as ordinary aspirin. This meant changing public attitudes to the drug, of course, because until then most British consumers hadn't even known that aspirin *was* irritating to the stomach. But although the strategy worked in the medium term (when given a choice, consumers will always go for the most palatable pharmaceutical product available), it was something of a double-edged sword. The more some people absorbed the idea that soluble aspirin might be preferable to ordinary aspirin because it had fewer side effects, the more others began to wonder whether it might not really be better to avoid the drug altogether. Aspirin began to get a bad name in Britain. Not surprisingly, then, when Panadol appeared as a *non*-aspirin alternative – an analgesic of apparently equivalent therapeutic value, but without any gastric problems – all aspirin sales, even those of Disprin, began to dip.*

Of course, Panadol's success also put Bayer Ltd in the anomalous position of selling an analgesic that posed a direct threat to its parent company's own aspirin products. This wasn't much of a problem in Britain because Bayer Aspirin sales were so low, but it caused some concern in New York. Eventually the matter was resolved when Sterling Products decided not to sell Panadol in the United States (a decision it would later regret). In the meantime, Panadol in Britain continued to go from strength to strength. Despite being sold as a prescription-only drug for the first fifteen years of its existence, it soon became highly popular with general practitioners. The decision to give it a memorable brand name helped. It was a trick right out of the old Carl Duisberg book of marketing: recognizing that busy doctors would neither remember nor bother to write N-acetyl-para-aminophenol or acetaminophen on a prescription pad when they could simply write Panadol instead. Indeed, the drug sold so well – and at such high prices – that the National Health Service became concerned at the effect it was having on its budget. In a blatant counter-attack, it took the unusual step of publishing the official pharmaceutical standards for it in the *London Gazette* in 1963 – a clear invitation to others to make it as a generic product. It even gave the drug a new name, paracetamol, in the hope that the medical fraternity would eventually remember it as well as aspirin.

As a long-term tactic this worked, but it had little immediate impact

* Although, as later became clear, the new drug wasn't without its problems either.

on Bayer Ltd's bottom line. Not long afterwards it began selling Panadol direct to the public as an over-the-counter medicine, and all those people who had once been prescribed it by their doctors now went to the chemist and bought it themselves. Other paracetamol products eventually appeared, as the NHS had hoped, but Panadol remained the market leader for years to come. Moreover, the cumulative effect on British analgesic sales was dramatic. By the early 1970s paracetamol had taken huge bites out of aspirin's market share in Britain and sales of the two drugs were almost on a par. The writing was on the wall for the world's favourite pill.

If competition was tough in Britain, it was as nothing compared to the situation in the United States. Sterling Products had been king of the American aspirin business before the war, maintaining its lead even when rivals launched attention-grabbing new products. But with all its energies focused on the furore over its relationship with IG Farben, it committed the cardinal sin of taking its domestic market for granted. What happened next was proof of the rule that pharmaceutical companies, like nature, abhor a vacuum. Sterling's competitors moved in for the kill.

Although there were hundreds of aspirin producers in America after the war, two in particular emerged to challenge Sterling's supremacy – American Home Products and Bristol Myers. The fight between these three would shape and define the US analgesic business for the next two decades. The ferocity with which they would batter each other – the claims and counter-claims, the smears and allegations, the vast sums of money spent on advertising and lawyers' fees, the inevitable intervention of the regulatory authorities, were all the consequence of the fact that the companies were essentially trying to sell the same thing to the same people. It was a battle in which the combatants had one aim, to convince consumers that *their* product was more effective than those of their rivals. The tactics were simple: identify some attractive or powerful property that only your aspirin had (or you could claim it had) and then advertise it vigorously, making absolutely sure in the process that your target audience got the message that it was a stronger, finer, more necessary and important quality than anything competing products might have to offer.

American Home Products was the first of the three to go down this route. The company had been set up in the mid-1920s – strangely enough, by Arthur Diebold, William Weiss's erstwhile partner at Sterling.* It took a beating during the Depression at the end of that decade but emerged as a leaner and meaner enterprise, determined to make the best of what was left. Its aspirin brand was Anacin, a product that had been around since 1930. The company's marketing pitch was based on the premise that the combination of the three ingredients Anacin contained somehow made it more powerful than plain aspirin, which was odd because aspirin was actually the dominant substance in its own product (a fact American Home was rather shy of admitting). It had always sold reasonably well because the company advertised aggressively but when it began to use television commercials in the early 1950s (still a novelty at the time) it really took off. The ads were crude but very effective – three hammers pounding away inside a man's head, each representing one of the symptoms that Anacin's unique combination formula was supposed to cure – worn nerves, tension and a thumping headache.

The next in the trinity was Bristol Myers, a former proprietary remedy firm. It made Bufferin, which like Anacin was mostly aspirin but had added antacids to speed up its absorption into the blood. (This was actually a genuine effect, but Bufferin's promotional slogan, 'Acts twice as fast as aspirin' was a little disingenuous because there was no proven correlation between speed of ingestion and speed of relief.) The company also owned Excedrin ('Four ingredients instead of one!') and so was able to fight on two fronts. It too embraced TV advertising and began to pour money into the new medium.

When Sterling Products finally got around to recognizing that the competition from these two rivals had cost it half its market share (and that they had grabbed some of the best slogans), it had little choice but to start fighting back with the selling proposition it had always used – that Bayer Aspirin was the pure product, untainted by any expensive and unnecessary additives. It joined in the orgy of television spending and began hammering home its message. 'Don't pay more for aspirin in disguise,' its commercials boomed.

---

* Diebold had a habit of buying small drug companies on the side in the hope that one day they might prove financially interesting. His stake in American Home did not survive the Depression.

Thus the three producers raised the banners under which (with a few variations) they would fight each other for the next thirty years: Power, Speed and Purity. Their rivalry would do more than just make a generation of American television viewers wearisomely familiar with images of silhouetted human heads, bubbling stomachs and intelligent-looking laboratory scientists with white coats and clipboards. It took competition to levels of intensity that the aspirin trade had never before seen, and, of course, made it extremely difficult for consumers to judge which product was better than the others.

Perhaps inevitably, regulators eventually began to point out that none of them was better. In 1962, the Federal Trade Commission, infuriated by this wild advertising, commissioned an independent study into the comparative efficacy of the leading brands. To the delight of Sterling's executives, it found that Bufferin and Anacin were actually no better and faster than 'ordinary' aspirin. The company immediately issued advertisements claiming that an official government medical team had revealed: 'Bayer Aspirin brings relief that is as fast, as strong and as gentle to the stomach as you can get.' But while this may have literally been the case, the claim outraged the FTC, which didn't like Sterling using its material in support of a commercial promotion. Unofficial war was declared.

Stung, the agency retired to its corner for a year or two and then came back with a demand that the aspirin companies must part with any scientific research they had supporting their advertising. If they couldn't unequivocally substantiate their claims, they shouldn't be making them. It followed this up by filing federal charges against the three major brand owners on the grounds that they had failed to let consumers know that medical opinion was divided about the nature of any real difference between them. American Home and Bristol Myers were separately cited for not revealing that Anacin and Bufferin were largely made of aspirin, and Sterling was charged with erroneously claiming that its aspirin was better than other brands. (It might be as good as the others, said the agency, but there was no clear proof that it was any better.)

Hundreds of advertisements were cited in support of these charges: ads that claimed that Bufferin 'acted faster on tension headaches' or was 'better for sensitive people', that Anacin was more powerful because of added caffeine, or that the Bayer brand was 'the world's best aspirin'.

The FTC made it plain that if any of these claims were found to be misleading it would insist that the companies spend sizeable sums of money on advertisements that put the record straight. What followed was years of legal arguments, submissions and preliminary hearings; a veritable tidal wave of documents in support of one contradictory position or another – all before the substantive matters even came to court.

In the meantime, of course, the commercial battle between the three brands continued unabated. But it wouldn't stay unchanged. Before long there was another major player on the scene. Sterling Products' decision to keep Panadol out of the US analgesic market had come back to haunt it – in the shape of a tiny red plastic fire engine.

This whimsical container was the brainchild of a small Pennsylvanian pharmaceutical company called McNeil Laboratories and held a new liquid brand of acetaminophen – the drug British consumers would come to know as paracetamol. McNeil had spotted a gap in the market for children's analgesics and, realizing that the drug was easily soluble and apparently free of side effects, dreamt up the idea of a toy-like dispenser in which to market it. The product was called Tylenol and following its launch as an ethical prescription drug in 1955, it soon found favour with American doctors who discovered that many of their younger patients found it easier to take than aspirin. Indeed, the response was so good that in 1958 McNeil won Food and Drug Administration permission to market a more traditional-looking Tylenol pill to adults as well.

In the meanwhile, the company had been taken over by a much larger business called Johnson and Johnson, famous for its Band-Aid plasters, baby powder and other assorted toiletries. At first it followed McNeil's low-key approach of selling Tylenol as a purely ethical prescription drug. After all, it was so effective. Within five years, without any consumer advertising at all, the product had become one of the 200 most commonly prescribed drugs in the United States.

But success like this didn't go unnoticed for long in the world's most competitive analgesic market, and when copycat acetaminophen products began to appear, Johnson and Johnson had to up the ante by investing in a major marketing campaign to cement Tylenol's position. In 1967, consumers were told that the drug they had hitherto been obtaining through their doctors was now available without prescription. Not long afterwards it launched an Extra Strength version (a slightly larger tablet) that was sold on the back of one of the cleverer sales pitches in marketing history: 'You can't buy a more potent pain reliever with a prescription.' Of course, all Johnson and Johnson was really doing was claiming parity with other products, i.e. that in non-prescription doses Tylenol was equally as powerful as anyone else's analgesic. But the overall impression given by the advertisements was that Tylenol was superior in some way. Sales shot through the roof and within a year it had become the biggest-selling analgesic brand in the country.

The leading aspirin makers were on the receiving end of this acetaminophen sales onslaught and as a result saw all the millions of dollars they had spent on promoting their own products (and on fighting among themselves) come to naught. Bayer Aspirin, whose owner, Sterling Products, had unaccountably stuck by its disastrous decision to keep its Panadol brand on the far side of the Atlantic, was one of the worst hit. Its share slumped to barely 10 per cent. To make matters worse, Tylenol's commercial success coincided with the court proceedings initiated several years earlier by the Federal Trade Commission. Over the next three years, Bristol Myers, American Home and Sterling Products were castigated for the misleading claims they had made in advertisements for their products. For all the hyperbole and the clever sloganeering, aspirin, it seemed, was just aspirin. It was a terrific drug, but there was nothing to choose between any of the leading brands. They were all the same, and that was official.

Then, just when it seemed as though aspirin's plight couldn't deteriorate any further, yet another analgesic was thrown into this competitive stew.

> We, Boots Pure Drug Company Limited, a British Company of Station Street, Nottingham, England, do hereby declare the invention, for which we pray that a patent may be granted to us, and the method by which it is to be performed, to be particularly described in and by the following statement . . .

On 12 January 1962, this elegant if impenetrable paragraph in UK Patent Specification 971700 introduced a novel class of therapeutic compounds to the world. It was the basis for a new drug – not a reworking of an old medicine as acetaminophen and aspirin had once been, but an original discovery, the first totally new analgesic and anti-inflammatory in several generations. Its inventors knew it by its tongue-tripping generic name, 2-(4-isobutylphenyl) propionic acid. To everybody else it was ibuprofen.

Boots is a familiar presence on British high streets today, where its general-purpose dispensing chemist shops sell everything from drugs and vitamin pills to sunglasses and sandwiches. But it's a company with a long history. It was founded in 1883 by Jesse Boot, an entrepreneurial Nottingham chemist, who quickly built the small family firm into one of the most successful retail chains in the country. On his death in 1921, its 600 stores passed into American hands for a few years when the company was merged into a broader US group known as Drug Inc. However, the enterprise failed during the Depression of the early 1930s and the Boots chain returned to British ownership. It then went back to doing what it did best, selling a huge range of products in an ever-expanding network of stores.

But Boots had always been more than just a dispensing chemist and general retailer; right from the start it manufactured medicines too (the company was one of the first to produce British-made aspirin). In the early 1950s, it set out to expand this area of the business and charged its small but well-established research department with the task of finding new pharmaceutical products. One of its early targets was a possible replacement for cortisone, an anti-inflammatory hormonal agent first isolated in the 1930s. The drug had proved particularly successful in treating joint inflammation caused by rheumatoid arthritis and had shown some analgesic potential as well, but unfortunately it could also have some very unpleasant side effects: ugly skin disorders, heart disease and stomach ulcers. Boots thought there was a market for a less injurious alternative. In 1954, it handed the task of finding it to Stewart Adams, a young pharmacologist with a doctorate from Leeds University.

In those days the company's medical researchers worked out of a large Victorian house adjacent to the main Boots headquarters in Nottingham. In what had once been a family living room Adams set to work reading all the available literature on the subject. He quickly became convinced that the anti-inflammatory properties of aspirin might prove the model he was looking for and that if he found a substance with a similar capacity it might also have aspirin's analgesic powers. But where to begin? Over the next two years he hunted through various chemical groups, ranging from all the known pain medications to substances that had originally been designed as weed-killers. But when he began to narrow the search down he encountered a problem. At the time there was no really effective way of testing compounds for their potential therapeutic activity. Without one it would be difficult to measure the performance of any discovery he made against that of existing drugs. So, somehow, he had to devise one of his own.

Salvation came when he read in a German journal about a method of inflaming the skin of guinea pigs by shaving strips of fur from their backs and subjecting the exposed flesh to ultraviolet light. It gave them a kind of sunburn and Adams realized that if he were able to induce it by quantifiable degrees, he would have a scale against which his experimental anti-inflammatory treatments could be measured.

The method worked, and with the help of a Boots biochemist called John Nicholson, Adams began putting his proto-drugs to the test. It was a massive and time-consuming task because there were so many potential permutations to try, and Nicholson had painstakingly to codify and prepare each one. But they began to get promising results. One potential drug (known as BTS8402) even made it as far as clinical testing because their guinea pig scale showed its anti-inflammatory effects to be six to ten times that of aspirin. Unfortunately, it had no effect whatsoever on humans. Further experiments, using a new American procedure for evaluating a drug's pain-killing effects on rats, showed that its analgesic and temperature-lowering properties were also very low. The latter was particularly important because Boots had now decided it wanted any new compound to be a fever reducer too. It was back to the drawing board. Then another group of chemicals, phenyl acetic acids, began to look hopeful. One of these, number 10335, gave particularly good results against rheumatoid arthritis but it too was dumped when it became clear that it induced a nasty rash. A variant, isobutyl phenyl acetic acid, was more successful and even made it briefly on to the market as ibufenac, but it was shown to cause liver damage in some people and was quickly withdrawn.*

This soul-destroying process went on for years. Adams and Nicholson

* Ibufenac remained on the market in Japan, however, because curiously these side effects were not apparent in Japanese people.

were aware that they were looking in the right area but most of the compounds they prepared were either too toxic or not powerful enough. Then in 1961, they finally came up with another subset of chemicals, phenylpropionic acids. The most promising wasn't the most biologically active of the group but 2-(4-isobutylphenyl) propionic acid did remarkably well in tests. Measured against a similar prescription-sized dose of aspirin, it was around twenty times more effective as an anti-inflammatory, had sixteen times the analgesic power, and between ten and twenty times the antipyretic effect. More importantly, once it had been patented, clinical trails showed that ibuprofen, as it was now called, showed no serious short-term side effects. It went on sale in the UK as a prescription drug, under the brand name Brufen, in 1969. After fifteen years' hard labour, the Boots team had found a drug that could potentially outperform every other analgesic product on the market.

After a shaky launch (the cautiously low dosage levels that Boots recommended on its introduction seemed to diminish the drug's effectiveness) Brufen did very well in Britain and soon began to be noticed in the United States too. In 1974, the Upjohn Company of Michigan bought non-exclusive US prescription rights and introduced an orange pill, Motrin, which soon proved very popular with the medical establishment, and after a few years Boots began to market its own US prescription version called Rufen as well. But the big breakthrough came when American Home and Bristol Myers acquired over-the-counter rights to the drug in 1984. Battered by their battles with Tylenol, they seized the opportunity with alacrity.

In the preceding ten years, aspirin had taken such a hammering from acetaminophen that the main American producers had temporarily lost the will to fight amongst themselves. With Tylenol firmly established as the market leader, all their efforts had been bent on trying to knock its maker, Johnson and Johnson, off its perch. But nothing seemed to shake it from the top. When the Food and Drug Administration announced that Tylenol should carry health warnings because it might cause liver damage, the aspirin makers discreetly celebrated until it became clear that its sales had barely dipped. When Tylenol was withdrawn from sale for a year in 1982 after someone injected packages of the drug with cyanide and killed eight people, they expressed shocked sympathy and set out to capitalize on its absence – only to see their hated rival regain its market lead soon after its return.

Their misery was compounded when in the early 1980s studies showed a link between aspirin and Reye's syndrome. This is an extremely rare condition affecting children whose immune systems have been attacked by a viral infection such as flu or chickenpox. However, researchers had found that aspirin could bring about the disease, and under pressure from regulators, makers of the children's version had to begin putting warning labels on their product.* Yet more people switched to Tylenol as a result.

But the appearance of ibuprofen gave American Home and Bristol Myers a chance finally to do to Tylenol what Johnson and Johnson had being doing to aspirin. Here was a new drug with a terrific pedigree, liked by doctors and with apparently unassailable medical credentials. It didn't matter much that at the light dosage levels allowed for over-the-counter sales, ibuprofen had little if any therapeutic advantage over acetaminophen, or even aspirin for that matter (non-prescription analgesics are actually remarkably similar in their power and effects); it was a successful new product and a weapon with which to fight back. Of course, the company that put it out into the market first would be in the strongest position to lead that attack. American Home and Bristol Myers raced to get it on to the shelves. American Home made it with some weeks to spare. Its Advil brand hit the shops in June 1984 and established a commanding lead before Bristol Myers's Nuprin could get a foothold. Within a year Advil had two thirds of the US ibuprofen market in the bag and had carved out a 5 per cent share of the overall analgesic business. It was a carbon copy of Tylenol's rise to prominence and sweet, sweet revenge.

Unfortunately, it didn't do much for aspirin sales either.

Needless to say, what had happened in Britain and the United States was repeated all over the developed world: the appearance of acetaminophen / paracetamol and then of ibuprofen split the analgesic market into three. Once-mighty aspirin, the king of the painkillers, had lost its crown and was now fighting a losing battle against the two newer pretenders. It still sold in huge quantities and was particularly popular in countries where people couldn't afford the expensively branded

* Aspirin is not now recommended for any child under the age of sixteen.

alternatives that were now cluttering the shelves of chemist stores in the West. But in the largest and most profitable markets its glory days seemed over. Ordinary old-fashioned aspirin seemed destined to go the same way as the rival analgesics and fever treatments it had so comprehensively crushed in its youth. If it had any future at all, then perhaps it was as a cheap ingredient in cold and flu remedies – a sad fate for a medicine that had once ruled the world.

And then, just when all seemed lost, someone got around to asking a simple question. How exactly does aspirin work? The answer would redeem the little white pill's fortunes once and for all. Aspirin's renaissance was about to begin.

11

SO THAT'S HOW IT WORKS!

IN LATE NOVEMBER 1956, the great and the good of America's pharmaceutical industry gathered at the headquarters of the Monsanto Chemical Company in St Louis, Missouri. They were there to attend a one-day symposium in celebration of an important event – the company's production of its 100 millionth pound of aspirin. Monsanto had been making the drug since 1917 (when Farbenfabriken Bayer lost its US patent) and over the intervening thirty-nine years had supplied it in raw form to many of the world's leading drug firms. It was a triumphant and complacent occasion; aspirin was still in its heyday, as yet untroubled by competition from newer analgesics like acetaminophen and ibuprofen, and only the year before had racked up almost $200 million in sales in the United States alone. Dr Carroll A. Hochwalt, Monsanto's head of research and development, gave the keynote address. He told his audience something of the known history of the drug, reminded them of its therapeutic benefits and reassured them (inaccurately, given what was coming) about aspirin's continued popularity. In conclusion, he turned to the drug's future:

> There is still much to be learned of the compound's mechanisms of action. How does it do what it does for the rheumatic and arthritis sufferers? Why does it reduce abnormal body temperatures while having no effect on normal temperatures? Just what is the means by which aspirin succeeds in moderating pain? The answers to these and other questions are sure to come in time and with them may come still more important applications for this cheapest, safest and most perennially durable of the genuine wonder drugs.

Dr Hochwalt was right on both counts. The answers would indeed come and with them his hoped for new applications, but it would all happen a lot slower than he imagined. Remarkably, given the vast amounts of money that had been made from the drug, the pharmaceutical industry had never been very interested in finding out how it worked. Nor had the medical establishment been bothered: as long as the drug functioned well enough and was relatively free of side effects, what else was there to say? In fact, while there had been a myriad studies on aspirin's benefits, no one since Bayer's Heinrich Dreser had actually bothered to investigate its mechanism – and he, as it later turned out, had got it all wrong. It would be another two decades before someone got around to unlocking aspirin's secrets. And when they did it would be as much by accident as design, one of those startling and unexpected discoveries that turned science on its head.

To the boy it was another exciting experiment, a chance to try out his new Bunsen burner. To his long-suffering parents it spelt disaster for their newly redecorated kitchen. History is a little unclear about what the young chemist was trying to do that day, and after all these years he can't quite remember himself (although he thinks it may have been a stink bomb), but the explosion that changed the colour of the freshly painted walls set him on the road to a Nobel Prize and an answer to the mysteries of aspirin.

It was wartime – Birmingham, England, 1940 – and thirteen-year-old John Vane was finding out that the course of science doesn't always run smooth. As life was full of unexpected surprises right then, he took it in his stride. A few months earlier, at the outset of war, his school had been evacuated to the country, returning only when the anticipated German bombing didn't materialize. Of course, the bombs then duly began to fall and John had already become accustomed to spending his nights with the rest of the family, huddled in the air raid shelter at the bottom of the garden. That may also have been why Mr Vane took such a pragmatic view of his son's experiments. After all, he'd given the boy the chemistry set in the first place, and what was another small explosion among so many? But kitchens were expensive, paint was hard to get hold of, and everyone's nerves demanded that something be done. Fortunately, Vane senior ran a small business making portable buildings and was able to erect a wooden shed near to the air raid shelter. Fitted with a bench, gas and water, it became John's first real laboratory, and he was able to make 'stinks and bangs' to his heart's content.

The hours he spent there fed a passion for chemistry that carried on through the next four years at Birmingham's King Edward VI High School; and when in 1944 the time came for him to leave, it seemed only natural that he should pursue it at the city's nearby university. But disillusionment soon set in because the undergraduate chemistry course seemed more about theory than experimentation – the thing John Vane loved most. 'The practical classes were devoid of any interest,' he said recently. 'They were boring because you were given a recipe. This is what you have to do. This is what you don't do. And the only excitement came when you got a 60 or 70 or 80 per cent yield. That was the glory, if you got a big yield. I hated it.'

Not surprisingly, when his professor, Maurice Stacey, asked him towards the end of the course what he wanted to do with his degree, Vane said, 'Anything but chemistry.' Stacey then told him he had received a letter that morning from a Professor Harold Burn at Oxford University asking whether he could recommend a young chemist to go to Oxford to be trained in pharmacology. Would he be interested? Like a drowning man clutching at an outstretched hand, Vane said he was – although the moment he left the office he had to rush to the library to find out what pharmacology was all about.

He arrived in Oxford in 1946 with little of the necessary biological training and even less motivation (his sojourn in the library hadn't been that illuminating). But Professor Burn turned out to be an inspiration. The school of pharmacology that he had founded before the war was already establishing a reputation as one of the most important research centres in Britain and Vane soon found himself in the company of some of the brightest and most impressionable scientists of his generation. Burn was a terrific teacher and stimulated and encouraged his young charges to unexpected heights. Most importantly, he continually emphasized the importance of experimentation and the need for pharmacologists to be constantly alert to the unusual. It was music to John Vane's ears. He had found his métier at last.

For the next two decades he lived a life of scientific achievement, moving smoothly through the stages of academia towards the pure research that is the prize of any ambitious scientist. His career took him first to Sheffield University as a laboratory worker, back to Oxford for his doctorate and marriage, then on to New Haven, Connecticut, and an assistant professorship at Yale University. In 1955, he returned to

England and a teaching post at London University's Institute of Basic Medical Sciences (then being run under the aegis of the Royal College of Surgeons). Before long he was its Professor of Experimental Pharmacology, a job that could have been designed with him in mind.

It was there that Harry Collier caught up with him.

Someone once said that 'In science the credit goes to the man who convinces the world, not to the man to whom the idea first occurs.' Arthur Eichengrün might have agreed with this aphorism, as indeed might Charles Gerhardt, the Strasbourg academic who first synthesized acetylsalicylic acid in 1853. But it's a glib saying that doesn't do justice to the fact that science is a community of ideas. For all the rivalry of individual researchers and their rush to get papers into *Nature* or *Scientific American*, they are usually the first to admit when their triumphs are built on the foundations of others; that without hundreds of years of hard-won technique and wisdom, or even, in many cases, the work of contemporaries in the same field, their discoveries would have been impossible. The story of aspirin is no different. At every stage of its development one piece of work had given rise to another and then another – a process that continues to this day. None the less, Harry Oswald Jackson Collier would always feel that at least some of the credit attaining to the momentous scientific feat of cracking aspirin's secrets had been denied him. As he would one day write to John Vane, 'If you prove to be, as probably you will, the Jesus Christ of aspirin, I think I may claim to be its John the Baptist.' It was the rationalization of a man who felt he had led a marathon from the beginning, only to be beaten to the tape by someone who had just joined the race.

Harry O. J. Collier was born in 1912 in Rio de Janeiro, where his father, a civil engineer who travelled the world building bridges, was working on a project for the Brazilian government. Unfortunately, the frequent upheavals the job required took their toll on Harry's parents' marriage and they separated when he was three, Harry returning to Britain with his mother. The rest of his childhood can't have been easy because money was tight, but he was a bright pupil at school and won a scholarship to Cambridge. There he excelled, taking a double first in zoology and chemistry, and began to acquire many of the wider interests that later made him such an engaging companion to his friends. His polymathy was extensive: he spoke several languages, loved literature and art, wrote scripts on science for radio documentaries and had a fine grasp of history to boot. But his first love would always be biochemistry.

In 1941, around the time that John Vane was exploring the delights of his garden shed laboratory, Harry Collier was taking his first tentative steps in the pharmaceutical business. He had a PhD and a teaching job at Manchester University behind him but without a large family income to fall back on he needed to earn more to support his wife and new family. Industry, though unglamorous in comparison to pure research, was far more lucrative, and because it was wartime there were plenty of openings with companies working on government contracts – Collier's first job was testing penicillin for the armed forces. But he soon moved to a medium-sized drug concern called Allen and Hanbury's (later bought by the pharmaceutical giant Glaxo) and from there to the London branch of Parke Davis, an American company from Detroit, where he set up its British pharmacology department and began to develop new drugs.

One of the first products he worked on was a synthetic version of poisonous curare, used as a muscle relaxant. The substance worked by interfering with the way nerves communicated with muscles. Collier became curious about this effect and began to broaden his researches into how pain signals were transmitted and prevented – a related process. It had recently been discovered that damaged cells release compounds called kinins into the blood, which attack adjacent nerve endings and induce inflammation and pain. To lessen pain, people took analgesics like aspirin but no one was quite sure exactly how they worked. Was it possible, he asked himself, that analgesics interfered with kinins in some way? Harry Collier's lifelong interest in aspirin was born.

In 1958, he began to experiment on guinea pigs to see if he could discover more about the way aspirin worked. As pain levels are notoriously difficult to quantify directly, he decided to administer a particular kinin, known as bradykinin. This causes a guinea pig's air passages to constrict – a much more obvious and measurable effect. Then he began giving aspirin, trying it before and after injecting the bradykinin. It was apparent that aspirin given *after* the event had no effect, the guinea pigs still experienced problems in breathing – bronchoconstriction, as it was known. But if given *before*, the air passages remained open and unaffected. Clearly, aspirin inhibited the bradykinin in some way.

Although intriguing and important, this discovery gave rise to another question. Did the aspirin act locally on the cells in the air tracts or on the central nervous system? There was only one way to find out. Collier repeated the experiments on guinea pigs whose vagus nerve had been cut, thereby severing communications between the animals' lungs and their brains. As earlier, the bradykinin caused the air tracts to close, although now the guinea pigs could not sense it happening. He then administered aspirin and once again he saw that it inhibited bradykinin's effects – the air passages remained open. The implication was clear: whatever aspirin was doing, it was doing it locally, not via the central nervous system. When the same experiments were tried using other basic analgesic substances such as salicylic acid it also became clear that aspirin was several times more potent in its effects. Taken together, these findings amounted to a vital breakthrough because they contradicted some fundamental beliefs about aspirin that had held sway since its invention.

Back in 1898, in the run-up to the drug's launch, Bayer's Heinrich Dreser had carried out some simple experiments to determine acetyl-salicylic acid's action on the body and to see whether, as Arthur Eichengrün had claimed, it was more palatable to the stomach than ordinary salicylic acid. He swallowed some aspirin powder and a short time later tested his urine. He found only traces of free salicylic acid. This must mean, he concluded, that the acetyl components of the drug (composed of carbon and hydrogen atoms) had dropped off in the stomach and that only the salicylic acid had passed through into the bloodstream. The salicylic acid must therefore be the *active* analgesic part of the drug and the acetyl group just an enabler that made the compound more digestible. Working on the basis of a common scientific assumption at the time – that analgesic drugs acted on the central nervous system rather than on local areas of pain – Dreser decided that aspirin's true benefits must lie in the way it allowed salicylic acid's analgesic power to get to where it was most needed – the nerve centres in the brain. He wrote this theory up in the article that accompanied the drug's launch, and it remained unchallenged until it was overturned by Harry Collier more than sixty years later.

As Collier's paper, published in the *British Journal of Pharmacology*, made clear, if aspirin was a more potent inhibitor of bradykinin than salicylic acid, it must also be the case that aspirin was more than just a palatable package for salicylic acid. It was obviously a powerful drug in its own right. Moreover, it clearly acted locally rather than on the central nervous system as Dreser had assumed. Although a great many questions were still unanswered, it was evident that everything scientists had hitherto believed about aspirin's action was wrong.

The more Collier worked on these experiments, the more excited he became. No one else was researching this particular field and he felt he might be on the verge of some even more interesting discoveries. Unfortunately, it wasn't easy to pursue them. Parke Davis was a profit-driven drug company. Collier's theories about aspirin, a drug already widely available, didn't seem to hold out any prospect of tangible returns and it was very difficult to persuade his bosses that his experiments were worth continuing. But surprisingly, he managed to bring them round. As his son Joseph Collier, now Professor of Clinical Pharmacology at St George's Hospital Medical School in London, said recently: 'I'm not really sure what they were doing, allowing him to carry on in that way. He was an industrial scientist not an academic researcher. Perhaps they thought that having one of their people doing some pure science lent them a bit of prestige.'

Whatever the case, Harry Collier was allowed to proceed, providing that any further experiments would be carried out on the margins of his other commitments. It can't have been ideal, but with the bit now firmly between his teeth he was determined to extend his theories. There was still so much to find out. As he was to write later:

> The questions that needed to be answered were: What local process did aspirin act upon and how did it modify that process? The answers to such questions came in a series of closer and closer approximations . . . One step in this approach to understanding the mechanism of aspirin action was the generalization that aspirin is an 'anti-defensive' drug. In other words, all the main bodily conditions for which aspirin is taken – fever, pain, inflammation – constitute parts of the body's system of defensive reactions. The benefits of aspirin could be attributed to the control of such defensive reactions, which had been misdirected or had become excessive.

But what Harry Collier didn't know was *how* aspirin controlled these defences. Did it interrupt them or did it make them more effective? It was an incredibly difficult and time-consuming question to answer because of the complex way that the body's biochemical processes unfold. These are analogous to a set of tumbling dominoes, where one action triggers a substance that releases a compound that stimulates another that precipitates a further reaction, and so on and so forth – all of them interrelated, but all fiendishly complicated processes that defy easy analysis. To employ another analogy, figuring out how they worked was like assembling a jigsaw puzzle while wearing a blindfold.

Fortunately, Collier was able to recruit some help. In 1963, Priscilla Piper was a bright young pharmacology researcher working for her PhD at the University of London. As part of her graduate programme she was seconded to Parke Davis to gain some experience of working in a commercial environment. After she'd spent a few weeks on his team, Collier recognized her as a kindred spirit and enlisted her support; she became his investigative assistant.

For the next five years the two of them devoted many of their waking hours to finding the answers Collier was looking for, trying hundreds of experiments on live guinea pigs, rats and rabbits. However, work that began in hope eventually became soul-destroying drudgery when they realised that they just didn't have the specialized skills to do the job. The only way really to determine aspirin's mechanism was to see the biochemical reactions with their own eyes. Working on live animals clearly made that impossible, but when they tried to remove guinea pig tissue in the aftermath of an experiment, the processes were disturbed and the effects they were trying to observe were lost. They had to get some expert advice.

It was then that Harry Collier thought of John Vane. The two had met through the Pharmacology Society and had become good friends – their families occasionally met up on camping holidays in the South of France. Harry's son Joe, then in the early stages of his career, was even spending two days a week conducting research in Vane's laboratory.

That laboratory, housed in the Royal College of Surgeons at Lincoln's Inn Fields in London, was providing John Vane with some of the most interesting and rewarding work of his career. He had attracted a talented team of graduate students and was already making a name for himself as one of the leading academic pharmacologists in the country. One of his notable achievements had been the development of a new way of conducting the bioassay. This is a complicated technique, distinctive to pharmacology, of determining the effects of chemical substances on animal tissue – obviously of huge importance to anyone wanting to test experimental pharmaceutical products. Vane's variation on this technique, known as the cascade super fusion bioassay, involved immersing strips of two pieces of tissue into a flowing neutral liquid called Krebs solution, one upstream from the other. The first tissue, usually guinea pig lung (perhaps from an animal previously given a dose of a particular test drug) would be injected with a substance drawn from egg white that made it go into anaphylactic shock – a dramatic and violent kind of seizure.* The damaged tissue would then secrete a hormonal compound into the solution, which would carry it over the second piece of tissue, usually from another animal such as a rabbit or a rat. If this reacted by twitching, flexing or changing in some other noticeable way, it was clear that it was responding to something in the hormonal secretion from the first guinea pig tissue. If you could pin down what exactly that substance was, you were some of the way towards determining the exact chemical reaction caused by the test drug.

Collier had heard of the technique and thought it might be useful if Priscilla Piper learnt how to do it. He asked Vane to take her on as a graduate student and teach her. As Vane was still in the process of refining his bioassay and needed an extra pair of hands, he was happy to agree, but his generous gesture turned out to have much greater consequences than either of them envisaged, because shortly after Piper switched camps they made a remarkable discovery.

They had been conducting a routine experiment with guinea pig lung. This had been sent into shock as usual and the resulting secretions were then carried by the neutral solution to a bank of six different animal tissues, including chicken rectum, rat stomach lining and rabbit aorta. Various chemicals were added to the torrent, each of which was known to neutralize one or other of the complex chemicals released by the guinea pig tissue. However, something must have been left unneutralized because for about thirty seconds Vane and Piper noticed the rabbit aorta twitching. Something new, a previously unidentified substance, had made it react.

What this chemical might be was a mystery: for want of something better they later called it rabbit-aorta contracting substance or RCS for

* Guinea pigs are violently allergic to egg white.

short (an abbreviation that raised a few smiles in the Royal College of Surgeons). But then Priscilla Piper, with five years of frustrating aspirin experiments behind her, suggested that they try injecting the drug into the guinea pig lung to see what happened. Vane hadn't taken much notice of Harry Collier's work up to that point, but saw no reason not to try this. To their astonishment, aspirin had a clear effect. If administered before the guinea pig lung was sent into shock nothing happened to the rabbit aorta. It remained unaffected. The drug had prevented the RCS from being released.

When he read their paper in *Nature*, Harry Collier was thrown into paroxysms of annoyance. He had lost Piper to John Vane and together they had used a technique to which he did not have access, to make deductions that he had hoped would be his. Obviously aspirin interfered with the new substance they had discovered. But how exactly, and what was RCS anyway? And what relationship did RCS have with Collier's own discovery that aspirin inhibited bradykinin? He sensed that the puzzle contained the final clue to aspirin's secrets. But who would decipher it first? The hunt for the true identity of RCS was on.

Among the many chemicals that Collier, Piper and Vane had used in their experiments was one group of hormone-like fatty acids called prostaglandins. These substances had been known to science since their discovery in the 1920s but because they are incredibly hard to detect in the body their complete categorization had been slow in coming. The most successful researcher in the field, the Swedish scientist Sune K. Bergstrom, had shown in the 1950s and early 1960s that they were all from a previously unknown class of chemical deriving from a slippery substance in living cell tissue called arachidonic acid. This makes our cells flexible and, in effect, allows our bodies to move. When something is done to stimulate or irritate cells (if they are injured, perhaps) they emit arachidonic acid and in the process unleash a slew of related chemical reactions. Out of that process prostaglandins are produced.

Other scientists had been intrigued by Bergstrom's work – any discovery of new organic chemicals always stimulates a rush to investigate their effects – and it emerged that prostaglandins were essential to the regulation of many important bodily functions, controlling everything from the elasticity of blood vessels and uterine contractions to the development of inflamed tissue around joints. It was also suspected that some prostaglandins produced effects similar to the ones seen in

Vane's various bioassays. Collier had even tried one common kind of prostaglandin in his experiments with guinea pigs, merely noting that aspirin failed to halt its effects. But John Vane was not so dismissive.

One weekend in April 1971 he was at home, tinkering with a forthcoming article that would summarize the work that he and Piper had managed to do so far. He hated writing papers at the best of times, but this was a particularly dispiriting project because almost two years had passed since their discovery of RCS and, despite many more experiments, they were no nearer to discovering its true identity. Then suddenly, with a flash of insight, a thought occurred to him. What if RCS was a prostaglandin, one that had not been previously identified? If that was the case, then aspirin was inhibiting a substance from one of the most important classes of chemicals in the body. What if the drug prevented the creation of other prostaglandins? Was that how aspirin worked?

It was a genuine Eureka moment – one of those rare intuitive leaps of genius by which the thorniest scientific problems are sometimes solved. If correct, the implications were startling. Aspirin was not the only analgesic around and several others had similar effects, ibuprofen being the most recent. Together they made up a whole class of drugs that reduced fever, pain and inflammation, known generally as non-steroidal anti-inflammatory drugs (non-steroidal because unlike substances like cortisone they didn't act like hormones), or NSAIDs. The mechanism of all these drugs was as much of a mystery as aspirin. If the famous little white pill inhibited prostaglandins, then was it possible that *all* the NSAIDs might act in the same way?

As he hurried into his laboratory the following Monday, these thoughts kept tumbling over one another in John Vane's head. There was only one way to find out if his theories were correct. He walked into the room and called for the attention of Priscilla Piper and his other colleagues. 'I think I know how aspirin works,' he said. 'I am going to do an experiment . . .'

Shaking off offers of help, he settled himself at a bench and began reading up on the techniques for preparing guinea pig tissue (as a professor he had previously left this tiresome task to assistants). Once he'd got the principles clear in his mind and after a few unsuccessful attempts, he managed to produce some that would pass muster. He put the samples into test tubes and shook them gently to provoke the production of arachidonic acid and prostaglandins. Then he repeated the process, only this time he also added aspirin. No prostaglandins appeared. Clearly, aspirin inhibited the production of one of the body's most significant chemical sets. It was the confirmation Vane was looking for.

But what did it mean? Think of a line of dominoes falling one on to another, each action precipitating another action. When cells are disturbed, they produce arachidonic acid. This leads to the formation of prostaglandins, which in their turn create fever or inflammation, and most likely (Vane reasoned) pain too. Then think of something coming between the first two dominoes and the others, preventing them from knocking over their fellows. That's what aspirin did. By stopping prostaglandins being made, it halted the resulting fever, inflammation and pain. John Vane had figured out how aspirin worked.

Ambitious though he was, Vane was also a decent man who realized how hard Harry Collier would take the news. He had cracked a mystery to which his old friend had devoted the previous ten years. But he knew, too, that there wasn't an awful lot he could say to comfort him.

Vane said recently,

> We went out for dinner. Harry was full of congratulations, although I knew that he was upset too. I think he would have liked it to have been his discovery. But I suppose in the end it also comes down to luck, or serendipity or chance. Call it whatever you like. But it plays a big part and the knack of being a good scientist is to recognize it when it comes along. To say, that's strange, that's funny and then to follow it up. I think I was lucky.

Joe Collier, who was working on his own unrelated projects in Vane's lab at the time, also remembers his father's distress. 'He was angry that he'd missed it, I think. But he would have buried that when talking to John, not wishing to steal his thunder.'

Vane was just as concerned about others finding out about the discovery before he was ready to publish. Joe Collier recalls that everyone on his team was immediately put on the project full-time and that although there was a great deal of suppressed excitement, the prevailing atmosphere was one of secrecy. Many more experiments had to be conducted to confirm the findings and the scientists were soon working long into the night. 'John could do that,' said Collier. 'There was no tension about it but he was very clearly in charge. He's always had this ability to follow things through when he has a goal in mind. He could turn an oil tanker around in mid ocean by sheer force of personality if he wanted to.'

On 23 June 1971, Vane and Priscilla Piper published the results in *Nature*, Britain's pre-eminent scientific journal. The article, entitled 'Inhibition of prostaglandin synthesis as a mechanism of action for aspirin-like drugs', would become one of the most famous and often quoted papers in scientific history.* It was accompanied by two related articles from other members of Vane's team that looked at the implications of the discovery. The applause was resounding.

In the years that followed a great deal more would come to be known about aspirin's mechanism and about prostaglandins generally (a field to which Vane would devote much of his subsequent career). Out of all this research would come an even greater understanding of the drug's effects. It was established that aspirin blocks the body's production of cyclooxygenase (COX), the enzyme that generates prostaglandins from arachidonic acid. Two distinct kinds of COX were identified (COX-1 that is involved, among other things in the formation of the protective lining of the digestive tract, and COX-2 that causes pain and inflammation) as well as several more types of prostaglandins, and inevitably, as this science became more sophisticated, it led to many more detailed ideas about the effects that all these complex biochemical processes have on the body.

However, for the purpose of this narrative it is enough to understand that aspirin is now known to have three principal modes of action, each relating to how much of it you take. In ordinary 300–600 mg doses, which people swallow for things like headache, it inhibits the prostaglandins responsible for the pain. Most headaches are caused by muscle contractions in the neck and scalp which stimulate the production of arachidonic acid. Aspirin prevents this being synthesized further.

In much larger prescription doses the drug acts on the swelling, heat and pain of inflammation – such as suffered by arthritis victims – and

* The paper also contained eight citations for Harry Collier's previous discoveries – a huge canon of work on aspirin going back to 1960.

again this is considered to be because aspirin inhibits the prostaglandin responsible, although some scientists have speculated that it might also interfere with the production of a kind of white blood cell called a neutrophil, part of the body's immune system. Under certain conditions rogue neutrophil can cause inflammation by attacking the proteins that make up human tissue.

The third action of aspirin is possibly the most important – its effect on the blood. There are three types of blood cells: red blood cells that carry oxygen from the lungs around the body; white blood cells that defend the body against infection and fight invasive bacteria; and platelets, part of the body's defence against bleeding. It is aspirin's interference with platelets that is one of its most remarkable effects.

Platelets are tiny flat discs, less than $^{1}/_{5,000}$ths of a millimetre across, and there are millions of them in every drop of blood.* They have a very short life cycle of around ten days and spend much of that time in a kind of limbo, floating around waiting for something to happen. They become active when they get a signal, usually triggered by arachidonic acid, which tells them a blood vessel has sprung a leak – or, in other words, started to bleed. Then they move very rapidly to the site of the damaged vessel and join together into a sticky mass that patches up the hole. This process is called platelet aggregation and is caused by a prostaglandin. In 1975, as part of the flurry of research that followed John Vane's discovery, Bengt Samuelsson (a Swedish scientist who had worked under Sune K. Bergstrom), identified this prostaglandin as thromboxane A2. Thromboxane is the main ingredient of RCS, the rabbit aorta-contracting substance that Vane and Piper had found six years earlier. By blocking the production of cyclooxygenase (COX), aspirin inhibits platelets from synthesizing thromboxane. It means they can't glue together, or clot, and start patching up a cut. The bleeding continues.

If you are someone to whom excessive bleeding is dangerous (someone like young Alexei Romanov, the haemophiliac heir to the doomed

---

* They were first described in 1842 by a French physiologist called Donne who mistakenly believed that they coalesced to become white blood cells. For many years other scientists dismissed them as cell debris with no useful function, but in 1874 Sir William Osler recognized the part that they might play in thrombotic processes. This theory would not be widely accepted until the late 1950s.

Imperial Russian throne), then clearly taking aspirin is a grave mistake, because it only makes your condition worse. But the aspirin–thromboxane relationship has its benefits too. Sometimes platelets can aggregate even when there is no cut. They can build up inside an intact but scratched or ulcerated blood vessel and cause what are known as thrombi – blockages that disrupt the blood flow around the body and clog vital arteries. The discovery that aspirin, in tiny 75 mg doses (all that's needed because thromboxane is so acutely sensitive to it) might prevent this happening was one of the great revelations of modern medicine.

It would have a number of consequences for the people behind the research and for the little white pill. In 1982, John Vane, who had gone on to find another prostaglandin, known as prostacylcin, which helps prevent blood clotting when the body is functioning healthily, shared the Nobel Prize for Medicine with Bengt Samuelsson and Sune Bergstrom. He would become one of the world's most celebrated scientists – honoured with a knighthood and awarded a fellowship in the Royal Society – and go on to play a part in the development of several important drugs for companies such as Wellcome. Now in the evening of his distinguished career, he runs the William Harvey Research Institute in London, one of the most important research establishments in Britain, and has finally, and a little reluctantly, given up the experimenting he always loved so much. His one-time assistant, Priscilla Piper, also became a celebrated scientist in her own right before her life was tragically cut short by cancer in the mid-1990s.

Sadly, Harry Collier was less fêted and it is only recently that his role in the unravelling of aspirin's mysteries has become more widely known. He left Parke Davis in 1969 and went to work for Miles Laboratories which, ironically, was taken over some years later by Bayer AG, the company from which aspirin had first emerged. He quietly continued his researches, however, and well into his supposed retirement he was working on experiments that might have established aspirin as a treatment for osteoporosis, the bone-weakening condition that affects the elderly. He died suddenly in August 1983 before his theories could be substantiated. A few days after his death, his son Joseph Collier delivered a lengthy paper his father had written on aspirin's history.

For the drug in question, the long-term repercussions of this research would be huge. It hadn't yielded up all its secrets (for instance, to this day it still is not known for certain why aspirin reduces high temperature but has no effect on normal temperature) but it had revealed a facet of aspirin's mechanism that would restore to it the sobriquet of 'wonderdrug' and offer its makers a chance of fighting back against the commercial onslaught of the new analgesic giants. It was the biochemical validation of a theory that had begun to excite doctors even before Vane's momentous discovery – proof, some said, that aspirin *must* prevent heart attacks, one of the biggest killers of the late twentieth century.

12

AFFAIRS OF THE HEART

A HUNDRED OR so years ago coronary disease was relatively rare. Thirty years ago just under a third of deaths in advanced nations could be attributed to heart disorders, and for every two people dying from a heart attack, another three suffered one. It is now widely believed that this astonishing seven-decade rise in coronary mortality was a consequence of twentieth-century prosperity; people's arteries clogged at a rate proportionate to their increased consumption of fatty foods, their lack of exercise, their smoking, drinking, stress levels and other risk factors associated with societies where it became all too easy, and affordable, to over-indulge and ignore the importance of personal health.* Fortunately, from the mid-1970s, as people became better educated about moderating their lifestyles and changing their habits, the mortality rate began to fall. But it still hasn't fallen far enough. Even today, more people in the Western world continue to die from heart disease than from any other cause and that doesn't even account for all those who succumb to strokes, a related condition. Obviously, then, anything that can help prevent those deaths is of the utmost importance.

---

* Widely believed but not universally accepted – at least as far as fat is concerned. In 2002, for instance, a Swedish epidemiologist, Uffe Ravnskov, challenged the long-held view that dietary fat and high cholesterol play a major role in causing coronary heart disease, citing the fact that several studies have found no significant difference in CHD rates between those with high cholesterol diets and those who eat leaner, low-fat foods. It's a complicated argument in which I do not feel qualified to take sides (hence I have repeated the established wisdom), but those who wish to know more (and perhaps feel a little happier about tucking into that fried breakfast) should get hold of Ravnskov's intriguing paper in volume 55 of the *Journal of Clinical Epidemiology*. It is perhaps worth mentioning, though, that fat intake in the Western world remained relatively high and stable in the 1980s and 1990s when CHD rates were falling most dramatically. Curious!

It took medicine a long time to catch up with the explosion in coronary disorders and to work out a response. The full story of how that happened is fascinating (if beyond the scope of this book), but suffice to say that from the moment that doctors began to devise diagnostic tools like the electrocardiograph (invented in 1903) and to understand more fully the causes and significance of hypertension, arteriosclerosis, angina, abnormalities in heart rhythm and other clinical signs of heart disease, they also began to think of ways of dealing with their consequences. Some of these responses would be in the realm of better public health, others would be surgical – bypass and other open-heart surgery techniques to relieve blocked arteries – but many would be medical, drug-based treatments based on an increased understanding of the way the circulatory system worked and how it could go wrong.

For many years the only drug of value for treating heart failure was digitalis, a preparation based on crushed foxglove leaves, first discovered by William Withering in the mid-eighteenth century, that slowed the pulse and helped relieve blood pressure.* But because it could be toxic, unpredictable in its effects and treated symptoms rather than causes, pharmacologists began looking for synthetic alternatives. Up to the end of the 1940s most of that attention was focused on a class of drugs called anticoagulants.

Coagulation is the name given to the process that takes place *after* platelets have rushed together to heal a damaged blood vessel, when a thick protective covering known as fibrin forms over the top of the platelets to seal them in place – similar to the scab on a graze. Scientists thought that if they could stop this happening it would prevent thrombi forming in diseased arteries. They began developing appropriate drugs – the anticoagulants. However, while some of these products turned out to be quite effective in treating *vein* thrombosis (versions of two of them, warfarin and heparin, are still widely used by heart surgeons today), it seemed that they were much less useful in treating blockages

* A medical botanist, Withering got the idea in 1775 from a Shropshire woman who used a herbal tea containing crushed foxglove to treat people with swollen legs. Unlike his near-contemporary the Reverend Edward Stone, whose ideas about willow bark were heard by the Royal Society but never acted upon, Withering managed to get his ideas taken seriously, and by the end of the eighteenth century digitalis was included in the *British Pharmacopoeia*.

in arteries where blood flow is faster. Platelet aggregation in damaged arteries takes place so quickly that life-threatening thrombi can form and do their worst well before slower-forming fibrin has had a chance to accumulate. In other words, there is no point using anticoagulating drugs to stop the production of fibrin in people with damaged arteries because by then it is too late – the patient has already suffered a heart attack.

The two Oxford scientists, John Poole and John French, who arrived at this conclusion were among the first to suggest that the best way to tackle heart attacks was to get to platelets before they started massing. Unfortunately, as their 1961 paper ruefully pointed out, 'Therapeutic measures designed to attack the platelet component of a thrombus are not yet available.'

Of course, that conclusion was not strictly true. But to be fair to Poole and French, no one had ever thought of aspirin as a possible treatment for heart disease – no one, that is, except Lawrence Craven of Glendale, California.

Craven was an MD but otherwise as far removed from the high-powered world of laboratory scientists and academic pharmacologists as it was possible to be. In 1950, he was working as a family doctor, with a side interest in ear, nose and throat cases, in a middle-class suburb of Los Angeles. It was an average sort of practice with modestly affluent and undemanding patients, but Craven was a bright and observant man who always sought to do his best for them and took note when his treatments seemed particularly effective or problematic.

Thus it was that in 1950 he wrote a paper for an obscure medical journal called the *Annals of Western Medicine and Surgery*, describing the effects of a post-operative treatment he had recommended to patients to relieve the pain of tonsillectomies. For several years he had prescribed a daily dose of four sticks of Aspergum, a mint-flavoured chewing gum impregnated with aspirin. But to his alarm he later found out that some of his patients had bled so profusely that they had to be hospitalized. On making enquiries, he found out that every person who had been haemorrhaging badly had gone out and bought additional packets of the gum and in the worst cases had been consuming up to twenty sticks a day – equivalent to around a dozen standard 300 mg aspirin tablets.

This had sparked his interest in the relationship between aspirin and bleeding to such an extent that he had begun to wonder whether the drug acted as an anti-clotting agent in some way and might even cut the chances of a heart attack. He was particularly intrigued by an apparent difference he noticed between the sexes: that women frequently used aspirin to alleviate minor aches and pains, while men 'hesitate to employ such allegedly effeminate methods'. Could this difference explain why men apparently suffered more from heart attacks than women – even though they might be equally old, overweight and unfit?

As Craven revealed in a subsequent paper for another minor medical publication, the *Mississippi Valley Medical Journal*, these deductions lay behind his decision to recommend a daily course of one or two aspirin tablets to all his patients and friends, many of whom had the kind of rich Californian lifestyle that would seem to mark them out for some kind of coronary event. He claimed that eventually he had persuaded 8,000 or so people to follow his advice, and much to his gratification, 'not a single case of detectable coronary or cerebral thrombosis occurred among patients who faithfully adhered to this regime during a period of eight years'. It was clear that 'aspirin administration offers a safe and sure method of prophylaxis against thrombosis'.

This was a groundbreaking discovery that might have saved many lives, but, tragically, the medical establishment failed to take it seriously. Craven was just a humble family doctor and he'd published his theories in obscure journals. Even worse, said his detractors, his methods were deeply flawed. He hadn't advanced any reason *why* aspirin should work in this way, nor had he followed any of the established rules of clinical trials, such as having a control group that took a placebo instead of aspirin. Why then should anyone pay any attention? Craven tried to fight back by publicizing his case in the popular newspapers, but to the big wheels in cardiology he was just an irritating crank who was way out of his depth.* It didn't help his case when he dropped dead in 1957 – from a heart attack. By the time Poole and French got

* Dr Craven's response to his critics was always to point to the living results of his experiments. It didn't matter that his trial was unscientific – his patients didn't have heart attacks and that was enough. As he told the *American Mercury* newspaper, 'I might answer that the mechanism whereby electroshock helps the confused is as yet unknown, yet few psychiatrists would discard electroshock treatment. Again, quinine is known as a specific for malaria – but can its worth be demonstrated by laboratory techniques?'

around to suggesting that a drug was needed to tackle platelet aggregation, Craven and his theories about aspirin had been long forgotten.

Fortunately, others *were* in a position to respond.

John O'Brien was born in November 1916 in London, to Australian parents (his father was a doctor, his mother a noted concert violinist). Educated at Westminster School, he won a scholarship to Oxford to study medicine and qualified in 1940. A man of boundless energy and enthusiasm, who played junior tennis at Wimbledon and later sailed in the fearsome Fastnet Yacht Race, he became a haematologist in Portsmouth, England, and devoted most of his working life to the study of thrombosis in vascular disease. Much of that interest came from early work he did on platelet stickiness and his invention in the early 1960s of a device called the platelet aggregometer, which both measured the cells' propensity to clump together and allowed scientists to judge how they were affected by other substances.* When O'Brien heard of Poole and French's work he began to use this device to search for chemicals that might stop platelets sticking. He tried a huge variety of possible drugs, from anti-malarial compounds to heroin and cocaine, and found that many of them worked, but regrettably only in doses that would kill the patient. And then, following up similar experiments being carried out by Harvey Weiss, a researcher from Mount Sinai Hospital in New York, he tried aspirin. It was a revelation – aspirin clearly had some anti-adhesive effect on platelets. He didn't know why (and no one would until the results of John Vane's researches were made known some years later) but its potential for tackling thrombosis was obvious. The question was, how to go about confirming it? Theories arrived at in the laboratory were useless unless they could be tested on real live patients. Someone would have to conduct a trial – a full-blown epidemiological study.

Broadly speaking, the main function of epidemiology is to examine the way a disease spreads through a group of people, ultimately with the aim of identifying patterns or links that can lead back to its cause and help identify an appropriate treatment. But many epidemiologists also like rolling up their sleeves and getting stuck into clinical trials

* Curiously, another version of the platelet aggregometer was invented around the same time by a man called Gustav Born – also a haematologist – who later become a close associate of John Vane.

– testing a potential remedy or therapeutic hypothesis on a sample group of people who suffer from a particular disease to see if that treatment is effective or not and what possible side effects it may have. One of the main sponsors of this work in Britain is a public body called the Medical Research Council (an adjunct to the NHS), which for more than seventy years has played a leading role in funding and organizing thousands of randomized controlled clinical trials – the gold standard for epidemiological study. Its work has encompassed everything from testing streptomycin on tuberculosis to investigating links between smoking and lung cancer, most famously established by Richard Doll and Austin Bradford Hill during an MRC trial in 1951. Naturally, when John O'Brien was looking around for someone to conduct a study on aspirin and thrombosis, it was to the MRC that he turned. He was eventually directed towards one of its regional centres in South Wales, where a bright young epidemiologist called Peter Elwood was making a name for himself.*

It took four years of repetitive house jobs – the most junior posts in the medical profession – and a six-month stint in general practice, before Peter Elwood decided he did not want to be a clinician. Born, raised and educated in Northern Ireland, he was a man who loved an intellectual challenge, and the prospect of seeing the same kinds of patient with the same kinds of complaint, day-in day-out for the next forty years, bored him rigid. Fortunately, during his time as a medical student he had taken some classes in epidemiology that fascinated him and as soon as he could he signed himself up on a research project that was examining the health of flax workers in and around Belfast. The work itself was standard stuff – looking at levels of tuberculosis, bronchitis and similar diseases and finding out if working conditions had contributed to them. But the process was gripping. 'I found it intellectually intoxicating,' he said recently. 'I used to walk home reading a textbook and often I would get so entranced I'd walk into a lamp-post.' He knew where his future lay.

In 1963, he had a chance to cement that future when he was invited by the MRC to join a new unit it had established in Cardiff, under

* The MRS did sponsor a small study that O'Brien ran on thrombosis but it was inconclusive.

the aegis of Archie Cochrane, one of Britain's leading epidemiologists. Cochrane was a pioneer of a kind of epidemiology that involved working in the community with large population samples and Elwood thought the technique offered intriguing possibilities for new areas of research. He began looking around for a project he could get his teeth into and thus it was he became interested in coronary thrombosis, and in particular the possible relationship between platelet aggregation and heart failure. He began to dream up a study that might evaluate the significance of platelets relative to other factors involved in cardiovascular disease and also look at what might determine platelet function – things like lifestyle, diet, exercise and smoking. If the study was successful, he might find clues as to what might help prevent heart attacks. While doing his background reading he frequently came across the work of people like John O'Brien and Harvey Weiss and was therefore delighted when the Portsmouth specialist contacted him through the MRC and suggested they co-operate. They began to correspond.

Elwood said:

> Our first face to face meeting actually took place in 1968, through a barred gate at the end of a platform in Portsmouth station. I was passing through there on my way to a conference in the Isle of Wight and I thought we could get together. But the ticket collector wouldn't let him through because he didn't have a ticket and for some reason wouldn't let me out, so we spent half an hour discussing the aspirin study through these bars. We must have been an odd sight.

The two men, it quickly turned out, were the answer to each other's prayers. O'Brien had been unable to persuade the MRC to fund a full trial that tested aspirin as a possible preventive for heart attacks because the sample that would be needed was too large. Around one in 200 people in Britain suffered annually from heart attacks, and to get enough people on whom aspirin could be tested would have meant signing up tens of thousands of volunteers. It was just too expensive. For his part, Elwood wanted to identify people with active platelets, because it seemed that active platelets were connected to heart attacks. Find the one and you might find a way to prevent the other. In discussion with O'Brien he realized that the aspirin might be used as a tool to reduce platelet activity. The patients who didn't take aspirin would be more likely to have active platelets and might therefore be easier to identify. The two researchers' aims fitted neatly together, just so long as they could get someone to pay for a study.

That is when Elwood had another clever idea. Older men who have survived one heart attack often have another – indeed, they are around twenty times more likely to have one than first-time (primary) victims. A trial that focused on just these secondary survivors meant they could reduce the size of the sample group needed and therefore make it likely that the MRC would grant funding. The MRC agreed, provided Elwood could persuade a drug company to provide all the necessary tablets – always a large part of the cost of a clinical trial. So he approached Nicholas Laboratories, the British descendant of the company that made Aspro and was responsible for some of the most extraordinary advertisements in the drug's history.

Things had changed at Nicholas. In the years after the war, the company had continued to export its unique marketing style around the world, selling Aspro in the boulevards of Paris, the jungles of South East Asia and the newly independent nations of North and Central Africa. But like many of the original free-wheeling aspirin producers, it had grown up and moved on to bigger things. George Davies, Nicholas's mercurial genius, was long gone and the business had evolved into a standard pharmaceutical concern with a wide range of products. It still made Aspro, though, and its executives were happy to meet Elwood's modest requests. He wanted them to supply aspirin and a matching placebo in specially made gelatin capsules, the idea being to disguise the bitter taste of the drug, which would make it impossible for the subjects of the study to know which they had taken.

Of course, setting up the trial was more than just a matter of getting the necessary pills. Elwood had to find enough subjects who had suffered a first heart attack and then persuade them to take part. Theoretically, he knew he should be able to find the patients by contacting local hospitals and asking cardiology specialists to tell him who had recently been admitted. But it all depended on getting his fellow doctors to co-operate.

> As it turned out, I had quite a few problems persuading people that it was a serious trial. They would laugh at me, and say: 'What on earth? Aspirin and heart disease. You must be kidding.' I used to take along all the papers and references and so on, and in the end most of them agreed because they

> thought it couldn't do any harm. But really, very few of them were convinced. They just didn't believe that this was a serious trial, that I wasn't having a hunch or playacting. Some of them even thought that aspirin might be a dangerous drug because they'd heard all these things about gastric side effects and bleeding. It was a little exasperating at times.

Eventually, though, he won enough of them around and in February 1971 the trial began. From then on it was a question of calling half a dozen local hospitals every Monday morning to get the names and addresses of recently discharged male heart-attack patients. Elwood would visit each of them at home, explain what lay behind the trial, and then persuade them to take one 300 mg aspirin dose a day. Most of the patients were receptive although some were a little concerned about possible side effects. Others would take a month's supply of capsules and then ask him, with a wink, what was really in them.

> They'd say, 'OK, doctor. I'll take them, but you're not going to tell me that it's aspirin!' They thought I was being deceptive and that it had a secret ingredient. Fortunately, I was a part-time lay Baptist preacher and that made it easier to look them in the eye and reassure them that I was being honest.

The trial was being run as a 'double-blind' study, which meant that neither Elwood nor the subjects themselves knew which of them were taking real aspirin and which were taking a placebo. The capsules all looked the same and came in identical packaging. Each small tin was number coded and another researcher, independent of Elwood, kept a note of the codes without knowing the fate of the patient. It was designed to rule out the possibility of bias, but it meant that while the trial was under way, Elwood had no means of knowing how it was going. With nothing else to do except recruit subjects and send them pills, he just had to wait patiently for results.

And then one Saturday morning, about a year into the trial, Elwood popped into the MRC offices in Cardiff's Richmond Road to pick up his mail. The phone rang. It was someone calling from Boston in the United States, in a state of barely suppressed excitement. Was that Dr Peter Elwood, and if so, was it true that he was conducting a trial on aspirin?

The caller was an American epidemiologist, Herschel Jick, who had been conducting a study of thousands of patients admitted to hospitals in and around Boston.

> They were questioning people on the fifth or sixth day after admission and asking them about the drugs they had taken a week before they were admitted. They were looking for unknown side effects and unknown beneficial effects and they hoped to be able to link the diagnosis with the drugs they were taking. They had forty diagnoses and around sixty drugs and had produced this great big chart matrix of all the correlations. The thing that stuck out in that whole matrix was that of the people who had taken aspirin in the week before coming into hospital, almost nobody appeared with myocardial infarction.

The figures were startling. On the one hand they appeared to show that people taking aspirin had an amazing 80 per cent lower chance of a heart attack – an almost unbelievable figure. But turn the analysis on its head and there was another more frightening possibility – that aspirin actually *increased* the chances of a heart attack being fatal. It might be that anyone taking aspirin had died before they had made it into hospital, or at least before they had been included in Jick's statistical evaluation. The absence of heart attacks among those of his subjects who *had* taken aspirin might simply be because none of them were likely heart attack victims in the first place. 'I could see his dilemma straight away,' said Elwood. 'Either aspirin was preventing heart attacks or it was killing them. And we were in the middle of a trial which was giving people aspirin!'

One of the Boston team was an English doctor from Oxford called Martin Vessey. He and Elwood had met a few times at conferences and Elwood had mentioned his aspirin study in passing. When the results from the American study began to emerge, Vessey told Jick of the Cardiff trial and they had decided that Elwood and the MRC must be warned. 'So I asked them what they wanted to do and Vessey said, "Look, we have to know if aspirin is beneficial or harmful. Will you break your code?"'

Breaking the code meant finding out which patients had taken aspirin and which had taken a placebo – something that was anathema to a scientist involved in a clinical trial because it could introduce bias and undermine the validity of the data. All Elwood's work would be wasted.

But if they didn't 'unblind', they would have to wait until the end of the trial to find out if they were needlessly killing their subjects. It put Elwood in an impossible situation.

> I explained that one of the conditions of our trial was that we had decided only to accept deaths in the results. In other words, if any of our subjects had an MI but didn't die, then we wouldn't include them in the figures. This was because we knew that aspirin reduces pain and we had to rule out the possibility that some people on aspirin might have an attack and not know it, which would distort the results. So I said to them, we've only had about seventeen deaths so far and we'll never prove anything either way with that. But in the end, of course, we didn't have any choice. We knew we couldn't face the public in a year's time if it turned out that we had been killing people with aspirin. We had to break the code.

That same day, with a heavy heart, Elwood got his team together and started unpicking the code. To his relief, of the seventeen subjects who had died, only six of them were in the aspirin group – the other eleven had taken placebos. One thing at least was clear: aspirin wasn't a killer. But unfortunately, the difference wasn't big enough to prove the Boston thesis about aspirin's benefits either. The sample was just too small. And now the whole study was tainted.

But Elwood also knew that they might be on to something. Taken together, the Boston study and the small but clear difference between his aspirin and placebo groups were indications that the drug might indeed be beneficial against heart attacks. With more resources and a bigger sample there was still a chance they might prove it definitively. A few weeks later, Elwood and his boss Archie Cochrane travelled down to London for a very private meeting at the MRC headquarters. The Boston team were also there, along with Sir Richard Doll, Britain's foremost epidemiologist and the man who had helped to establish a link between smoking and cancer. The doors were locked to prevent any leaks to the outside world. For several hours they discussed what to do next.

Elwood said:

> It was all very exciting. Remember, in those days there was nothing you could do to reduce the risk of heart attacks. You could talk a man into

stopping smoking and tell him to take exercise, but really there was little medically you could do. Beta blockers [drugs that steady the heartbeat] hadn't yet properly come on to the market and here was something readily available that might be a real breakthrough. But we had to be careful.

The problem was, there was no way of knowing for certain. The Boston numbers could yet turn out to be an aberration, a statistical fluke, and tempted though Henschel Jick and his colleagues were to release them to the public and bask in the headlines, Elwood's early data was at best equivocal and might ultimately contradict them altogether. So the meeting arrived at a consensus. Elwood's study would continue, but with a big increase in resources to enable him to expand the sample. That way, the final results should be more conclusive. In the meantime, the Boston team would hang on to their results on condition that Sir Richard Doll monitor Elwood's study independently and would stop it if a strong negative or beneficial effect became apparent.

And so the British trial went on, expanded to include heart patients in Swansea, Oxford, Birmingham and Manchester and now exclusively focused on determining whether aspirin was as good as everyone hoped (Elwood's original ideas about platelet aggregation were put on a back burner, although he would return to them later in his career). In the meantime, Nicholas Laboratories produced more gelatine capsules and Elwood went back to his recruitment drive. With the benefit of hindsight, he now confesses to not having been ambitious enough. 'I suppose part of me still wanted to keep it under my control. But of course we didn't know then just how large a sample we needed.'

By the time the trial ended in September 1973, Elwood had signed up 1,239 male patients. It was a big number by the standards of clinical trials of that time, but as it turned out, not quite big enough. In the thirty-month period of the study, 108 men died; forty-seven of them had taken aspirin, sixty-one had taken the placebo. Aspirin appeared to have some benefit (around a 24 per cent reduction in deaths) but the difference wasn't statistically significant. The trial had been too small to prove the matter beyond doubt. When it was published alongside the Boston results in 1974, it generated quite a bit of publicity as it was the first evidence from a randomized trial that a reduction in cardiovascular disease might be possible, but everyone privately understood that another study would have to be carried out. The *British Medical Journal* ran it under the advisory headline 'for debate', which seemed to sum up what people thought of it – somewhat to Elwood's frustration.

> Many clinicians were not persuaded. Here was a trial showing some unexpected results but they weren't as conclusive as we had hoped. Every now and then I meet the occasional physician who tells me he saw that paper and started giving aspirin to his patients from that moment, but most thought, 'Oh well, they're bound to find it harmful again next year. We'll wait and see . . .'

So he went back to his office and drew up plans for another slightly larger trial, known as Cardiff 2. This time it took four years but again produced inconclusive results, aspirin apparently reducing heart attacks by around 25 per cent. It was a sizeable number but still within the possible margins of error and chance and not statistically definitive. Elwood got permission from his bosses at the MRC to try one last time, but now he asked doctors to give aspirin or a placebo, they didn't know which, to patients *during* a heart attack to see if the time of ingestion made any difference. It appeared to be totally ineffective – a result that left Elwood flummoxed. He had tried three times to show whether aspirin had any effect on heart attacks. On balance, the results had been highly promising, though sadly not yet conclusive. He had to call it a day.

In the meanwhile, however, the implications of the discoveries made by John Vane and Bengt Samuelsson, the Swedish prostaglandin expert who identified thromboxane, had begun to filter out. If aspirin prevented clotting then it *must* have an effect on heart attacks. Other epidemiologists took up where Elwood had left off, starting trials of their own in France, Germany and the United States. Again the samples were small and again the results were promising but inconclusive. It was tantalizing and deeply frustrating to all concerned. There seemed to be only one solution – put together a study big enough to rule out the effects that chance might be having on the figures.

The result was the Aspirin Myocardial Infarction Study (AMIS), which was begun in the United States in 1975 by the US National Heart, Lung and Blood Institute. Costing more than $17 million and lasting four years, it analysed the effects of aspirin on 4,524 heart attack victims (a sample that was more than twice the size of Elwood's biggest trial). When the results were published in 1980, they were deeply disappointing. On the positive side, those who had taken aspirin suffered 30 per cent fewer *non-fatal* heart attacks (a large number, although by the conventions of statistical analysis still not large enough to rule out chance). But significantly there were slightly more *deaths* in the aspirin group than in the placebo group. Even allowing for possible flaws, said the *Journal of the American Medical Association*, it wasn't enough. 'In summation, based on AMIS results, aspirin is not recommended for routine use in patients who have survived a heart attack.'

To the despair of aspirin's champions, that conclusion seemed to have settled the matter. But their unhappiness was shortlived. Two months later it was dramatically overturned.

Imagine for a moment that you are trying your hand at being an epidemiologist. As an experiment, you decide to re-create one of the early aspirin trials to see if you can improve on the results – to show, for instance, that the drug is beneficial in preventing second heart attacks. To make it authentic, you start from scratch and rely only on information that was available to people like Peter Elwood in the 1970s. From your reading you know that you have to take some basic things into account. The first is that the published research of the time suggests that aspirin might reduce secondary heart attacks by about a quarter, but it is not yet proven conclusively. From other contemporary statistics, you discover that around 10 per cent of people who have a first heart attack will have a second one within a year. Armed with this data, you then decide to run your trial over twelve months with 2,000 patients – all that your budget can afford (it's an expensive business, signing up all those volunteers, arranging supplies of pills and setting up an office to run things). Your plan is to give half the subjects aspirin (the treatment group) and the other half a placebo (the control group). If the coronary statistics are correct, at the end of the year around 100, or 10 per cent, of the control group will have had a second heart attack. Likewise, if all goes well and aspirin does what it's supposed to do, then of the 1,000 in the treatment group, only seventy-five will have had a heart attack. Bingo! You will have arrived at a nice neat conclusion. Twenty-five per cent of everyone who has had a heart attack will not have another one if they take aspirin. Already you can see yourself writing up your results for a prominent journal and sitting back awaiting the applause.

Oh, if life were that simple. Because the moment your trial has come to an end and you start compiling the results, it becomes clear that you have hit a number of snags. Your results aren't quite so clear-cut. Perhaps your subjects haven't behaved as the statistics said they would – maybe more of them in the treatment group have died than you anticipated. But is that because aspirin hasn't been as effective as you hoped, or is it because some of the subjects were much sicker than the statistics said they would be? Or what if, by some fluke, all of the 10 per cent of people who were destined to have second heart attacks had ended up in your treatment group rather than the placebo set? That might explain why you have had fifty heart attacks in the former and none in the latter. But how can you be sure? It might just be that there have been no heart attacks in the placebo group because none of them had taken the aspirin, and then where does that leave you – concluding that aspirin causes heart attacks? And, of course, the figures might have been further distorted by the fact that some of the patients have told you they've taken the pills as instructed, when in reality they've been flushing them down the toilet. You've monitored the subjects as best you can, but what if one or two rogue results had got through the system without you noticing?

Suddenly your confidence in your trial starts to slip.

That's when you begin to get an insight into the kinds of complication professional epidemiologists have to contend with. Every study runs the risk of being hit by statistical anomalies, particularly when the disease afflicting your patients is rare and the mathematical chances of everything not going to plan are increased. It's why medicine generally accepts that a single study is conclusive only when the results are clear enough one way or another to rule out fluke or error.* Moderate benefits of around 25 per cent can't be relied on unless the trial has been so huge that the chances of distortion have been reduced to a negligible degree. Even if your study *had* shown that aspirin reduced second heart attacks by 25 per cent, it could never be called conclusive

---

* Although this has never seemed to stop authors of small clinical trials of this or that new miracle treatment claiming in the media that *their* trial is the conclusive one.

because your sample was too small. A difference of twenty-five deaths between the control group and the placebo group, out of a total number of 2,000 subjects, is just not enough. On the face of it, the only way to get around the problem is either to run a trial over a great many years in order to get enough subjects to produce acceptable numbers (difficult in itself, because many of your volunteers might die from other unrelated causes along the way) or to come up with a massive sample right from the start. And both those options, of course, are monstrously expensive.

With the benefit of thirty years of hindsight, it is easier to understand how such problems affected the work and thinking of those involved in the first aspirin/heart attack trials. They kept on producing results that were promising but never quite clear enough to pass muster. The AMIS trial had seemed big enough to clear the matter up once and for all, but once its results came out, many epidemiologists like Peter Elwood knew that it just didn't feel right. It had nothing to do with hurt professional pride. Several aspirin studies leant towards the conclusion that aspirin prevented second heart attacks. One did not. Surely there was something wrong somewhere?

And then, in May 1980, an editorial appeared in the *Lancet*. It was based on the conclusions of a new professional association, the Society for Clinical Trials, which had just held its first conference. At that meeting someone had presented a new and very controversial way of looking at the aspirin problem – a mathematical procedure for evaluating all the clinical trials together. This demonstrated beyond reasonable doubt that the trials *collectively* showed one thing: aspirin helped prevent second heart attacks. AMIS had got it wrong after all.

The man behind this radical reassessment was one Richard Peto, an idiosyncratic Oxford statistician who had cut his teeth in epidemiology by working alongside Sir Richard Doll. Since the mid-1970s Peto had been developing a theory that explained why clinical trials needn't be seen in isolation – you could add them together and look at the big picture they painted. The mathematics behind it was a little complicated but it was a formula that ironed out the positive and negative effects that chance might have played on the results. Over a number of trials these would tend to balance each other out, making the results that much more reliable. Even though you might have a number of trials with different quantitative results, if they all pointed in the same direction you could treat the overall trend with some confidence.

The procedure, known as meta-analysis, was controversial because clinical trials usually have different methodology, featuring patients of different ages and gender, with different diagnoses. Some clinical researchers believed that lumping them together was contravening a fundamental rule of statistics: never try to pool raw data because you will introduce bias into your equations. But Peto was dismissive of this objection because he never had any intention of pooling data in this way. He was just interested in the answer to a big, simple question. Did the trials show that the treatment did any good? Yes or no?

Like many others in epidemiology throughout the 1970s, Peto had been monitoring the various aspirin studies and had arrived at much the same conclusion: that the only way to find out conclusively if aspirin prevented heart attacks was to conduct a really big and very simple trial. In 1978, with Sir Richard Doll and others, he set up a study involving more than 5,000 British doctors. (It was originally aimed at every male physician on the British medical register but thousands had to be excluded because they were already taking aspirin or had a history of heart trouble, stroke or stomach ulcers.) Two thirds of the doctors were asked to take aspirin, the rest were asked to avoid aspirin for the duration of the trial, taking paracetamol for headaches instead. Every few months they would then fill out a questionnaire on the state of their health. Even as they were putting it together, Peto and his colleagues were aware that the trial probably wouldn't be big enough, but their hope was that they would shortly be able to extend it to US physicians, giving them a sample of some tens of thousands.

But in the meantime the AMIS study was published, with its conclusion that aspirin did nothing much to prevent second heart attacks, and the chances of getting American help receded.* Like Peter Elwood

---

* The US National Institutes of Health, which the British asked for help, was also concerned about the apparent laxness of the supervision in the trial, believing that it might invalidate any conclusions. The British team argued that doctors could be trusted to self-medicate and report their health accurately and in any case, as the study was only meant to answer one simple question – does aspirin prevent heart attacks? – the sheer size of the sample would give a clear enough result. But to no avail. Their proposal was turned down, although the idea of a US physicians' trial would later be resurrected, with astonishing results (see below).

and others, Peto saw the AMIS result as an aberration – and something of a challenge. He decided to put it under the microscope of his meta-analysis with the five previous aspirin trials (two from the Cardiff team and others from Germany and France). The result was clear. AMIS did little to alter the overall picture. On the basis of that conclusion, he prepared a paper for the Society for Clinical Trials, which was about to hold its plenary session in Philadelphia.

To many clinicians (several of whom were already tentatively recommending the drug to heart patients under their care), Peto's widely reported presentation was the news they had been waiting for – sufficient confirmation that aspirin prevented secondary heart attacks in around 25 per cent of cases. The fact that the *Lancet* was also convinced swung many waverers: it was difficult to remain sceptical when an editorial in one of the world's leading medical journals was recommending that drug companies market aspirin in calender packs to make it easier for patients to remember to take them.

But there was still one more hurdle to overcome. The US medical authorities remained sceptical. The Food and Drug Administration (FDA), in particular, was hostile to the whole idea of meta-analysis and doggedly held on to the view that the only outcome of combining several small inconclusive trials would be one larger inconclusive result (an opinion not unrelated to the fact that its sister organization, the National Institutes of Heath had put up much of the money for the AMIS trial). And, of course, the FDA's opinion mattered because without the agency's backing it was impossible for a pharmaceutical company to market a drug in the United States, or advise consumers that an old drug had a new use. Anyone who wanted to market aspirin as a preventive treatment for secondary heart attacks had to get FDA approval. In 1983, one of the aspirin companies set out to win the agency around.

Sterling Products had been badly battered in the analgesic battles of the preceding three decades. From a position of almost total dominance, it had seen Bayer Aspirin's share of the US market slip to around 6 per cent by the early 1980s. It might have held its own against rival aspirin brands like Anacin and Bufferin, but the appearance of acetaminophen/Tylenol and then ibuprofen drugs like Advil had pushed it out on to the margins. Aspirin had seemed destined for a long, slow retreat into pharmaceutical oblivion.

And then came the scientific reports on aspirin's action and a succession of clinical trials that seemed to suggest that the little white pill might enjoy a renaissance as a heart attack drug. It was sweetly exciting news and every few months seemed to bring more. But how best to capitalize on it? How best to let the drug-buying public know what was going on?

Sterling decided it should change the labels on Bayer Aspirin packets. In December 1980, it applied to the FDA for permission to add a line to the promotional material it sent out to health professionals about aspirin (a first step towards changing the labelling on its product packaging). The sentence it wanted to include was simple enough: 'Aspirin has been shown to be effective in reducing the risk of death or re-infarction of patients who have suffered a myocardial infarction.' But unfortunately for Sterling, the AMIS study had been published some months before and the FDA felt there were no grounds to accede to the request. The company lobbied hard, but for three years the agency held firm. There would be no change in the labelling.

But then, in March 1983, Sterling finally managed to persuade the agency to hold a hearing of its Cardiovascular and Renal Drugs Advisory Committee, one of the many panels of independent experts that advise the FDA on the legitimacy or otherwise of claims made by drug firms. Naturally, the panel asked that Sterling accompany its application with a presentation of the relevant scientific data. As the company hadn't taken part in any of the aspirin trials, it had to bring in outside experts and asked Peter Elwood and Richard Peto to be two of its principal witnesses. Although reluctant to be thought to be at the beck and call of a large international drug business, they agreed (whatever Sterling's commercial motives might be, it was clearly important to convince the Americans of aspirin's benefits) and flew over to Washington. Unfortunately, from then on things didn't quite go to plan.

The first witness was Jack Hirsch, a heart specialist from Ontario in Canada. He ran though the basic medical science – how a substance called thromboxane led to platelet aggregation and how this in turn led to the formation of thrombi in arteries. Thrombi, he said, were the major cause of myocardial infarction. It had been discovered that aspirin inhibited thromboxane and it made sense that it should stop thrombi and heart attacks too.

Then Peter Elwood took over the podium and talked about the six major aspirin trials. With the exception of AMIS, which in his opinion was at odds with the facts, they all showed that aspirin had a beneficial effect on heart attacks. It was true that none of the studies, including his own, was statistically conclusive and it was right that they should be considered in that light, but the direction of the trials was clear.

Richard Peto was next.

Elwood said recently:

> You have to understand that Richard is a one off. He's a very bright man who can be, oh, how shall I put this, a little abrasive when people don't get the picture. And the atmosphere in the room was quite tense. The Sterling executive who was sitting next to us had told me that there was this tremendous campaign to rubbish aspirin and I think that affected things.

Dressed in his customary brown corduroy jacket, with no tie and with longish blond hair falling over his collar, Peto cut an unusual figure in a room full of conservatively suited executives. He clearly anticipated that his ideas were going to be challenged and he was in no mood to be patronized by an American panel that he suspected held unflattering views of British science. He ran quickly through his presentation on the principles of meta-analysis and how he had applied it to aspirin. Every few minutes he'd scribble out a chart before slapping it down on to an overhead projector. His *ad hoc* style was more informal than the panel members were used to and his ironic tone seemed to suggest that if the Americans couldn't see what he was on about, it was their failing not his. Suspecting that he wasn't taking the meeting seriously, some on the board, which included representatives close to Sterling's rival drug companies, began looking at him askance. In the questions that followed they let their petulance show.

The initial target was Peter Elwood's first trial. It had been 'unblinded' well before it was finished – how could he be sure that it hadn't been tainted? How certain was he that all his subjects had taken their allotted pills? Could there have been some failure in the randomization process? As one hostile question followed another, Elwood quickly grew exasperated.

> They really attacked me and I have to say it made me very angry. Someone would get up and say how do you know for certain that the men didn't just dump the tablets down the toilet. And so I said, well, I don't believe that's what happened because the men seemed to me to be perfectly honest – and then they'd interrupt and say well we can't just take that on trust. Anyway, they made out that I was an idiot and that my trial couldn't be depended on. I really felt that this was foolish. I didn't show it, but I got very angry. I mean, if you treat every trial like that there'll be no advance on anything . . .

The panel were just as dismissive about Peto's analysis. One board member, Phillip Dern, said that the evidence in favour of aspirin was flawed and that all the mathematical fiddling in the world couldn't hide the fact. Clearly, Elwood's trial had been broken too early and clearly, its data was suspect. Peto's figures were based on discredited and invalid material. This infuriated the statistician who knew that Elwood had opened his trial code only because some American researchers had asked him to. As Elwood recalls, Peto then started to argue back, his annoyance plain for all to see.

> Richard is not nearly as tolerant as I am and I remember we were sitting down and he was close to me and I could hear him saying, 'Rubbish. The man's a fool. He's an idiot. How can he ask that?' All in a very loud whisper that was audible across the room. I can see now that it was probably quite amusing, but of course the issue was very serious at the time . . .

Things had degenerated into a squabble, and not surprisingly when it eventually came to a vote the committee declined to accept Sterling's application. They would reconsider the matter only if some documentation was provided, which proved that the unbinding of Elwood's trial hadn't affected the decisions of the doctors diagnosing second heart attacks among the study's subjects. It had to show that they had no idea whether they were seeing someone from the control group or the treatment group.

The meeting had been a disaster for Sterling. What it had hoped would be a relatively straightforward process had turned into a hostile slanging match. Moreover, most of the attention had been on Elwood's trial and Peto's essential point about aspirin and meta-analysis had been lost. But at least another hearing hadn't been ruled out.

It would not be held for almost two years. In the meantime, prompted by Sterling, Elwood reluctantly went back over his data and wrote out a lengthy summation of how his trial had been run, presenting clear evidence to show that he had no idea of which of his subjects had received aspirin and which a placebo when he drew up his original results. Richard Peto, for his part, swallowed his anger at the FDA and put every single detail of every patient in the six aspirin trials through the massive Cray mainframe computer at Oxford University – some 10,000 pieces of data. In the process he found that the AMIS trial hadn't been as flawed as some had suspected – just desperately unlucky with statistical anomalies. But it made little difference to his overall analysis. The trend in aspirin's favour was unaffected.

On 11 December 1984, the committee reconvened. They heard first about a recent study conducted among 1,200 US armed forces veterans, which had tested aspirin on patients with unstable angina (a clear clinical sign of impending heart attack) over the course of a year. The aspirin takers suffered 43 per cent fewer deaths than the placebo group – by anyone's standards a big enough difference to be accepted as conclusive. It was a good start.

Then Peto came to the podium. Formally dressed and pre-equipped with slides and a written text, he gave a remarkably clear and effective presentation, a devastatingly thorough explanation of why meta-analysis could be applied to the aspirin/heart attack trials and why it was important to do so, even though single studies had proved inconclusive and had shown only moderate benefits.

> The clinical viewpoint on moderate risk reductions is that small reductions in the risk of death are not noticeable in ordinary clinical practice . . . You will hear some doctor saying, well, if it doesn't show up in a trial with a couple of hundred patients then it can't be worth bothering with. This is not medical wisdom but statistical unwisdom. The public health viewpoint is quite different. It says that worldwide the total effects of quite moderate risk reductions may be substantial. I mean, every year you are going to get about a million or so patients admitted to acute coronary units around the well-doctored world . . . About 150,000 of these will be dead within a year. Now, if you can knock about 15 per cent down to about 13.5 per cent – this is just a 10 per cent risk reduction – you are talking about prevention of 15,000 deaths a year . . .

> Some of these people will be old, some will be horrible people who would be better dead anyway, but a fair number of these are going to be in middle age with a reasonable chance of enjoying life. So this kind of thing is worth doing. You may not know whom you saved but nevertheless this kind of thing is worth recognizing. And at the moment we are missing them.

He then turned specifically to aspirin and step by step showed the committee how his meta-analysis worked and how when you mathematically calculated the positive and negative effects of chance on each study, and then removed them from the equation, the overall indications were that aspirin improved the odds of second heart attack death by between a fifth and a quarter.

It was a bravura performance. At the end of his presentation there was a brief silence and then the room burst into a spontaneous round of applause. Shortly after, when a vote was taken, the committee decided unanimously to recommend a change in the labelling. Supported by the data from the veterans' angina study (a useful fig leaf for anyone still claiming to feel uneasy about meta-analysis), Richard Peto had brought them round.

Just under a year later, the US Health Secretary, Margaret Heckler, held up a bottle of aspirin at a packed press conference and told the world's media that one tablet a day could prevent secondary heart attacks. The wonder drug was back in action.

13

A TWENTY-FIRST-CENTURY WONDER DRUG

ON 6 MARCH 1999, fifty mountaineers abseiled down from a 400-foot-high tower block on the banks of the Rhine, unleashing vast rolls of cloth that were tethered to scaffolding on the roof. As the material unfurled – all 242,190 square feet of it – those watching below burst into applause. One hundred years to the day since it first registered its most important trademark at the Imperial Patent Office in Berlin, Bayer AG was transforming its Leverkusen headquarters into the world's largest aspirin packet.

Although the anniversary was the principal cause of celebration, this was more than just a big birthday party. The company was also making a bold, even triumphant, public statement. Its most famous product was enjoying a renaissance as a potential treatment for some of the West's deadliest diseases, and Bayer – at last – had won back the rights to sell it in the most important market in the world. Lest anyone should doubt it, all of Bayer Aspirin was now back in the hands of its rightful owners. Older employees with an eye for symbolism might have spotted another, subtler, subtext of the ceremony as well. The building stood on a plot once occupied by Carl Duisberg's palatial villa, from where the company's guiding spirit had plotted aspirin's spectacular rise so many years before. How he would have savoured this moment.

Bayer AG had come far since its emergence from the ashes of IG Farben. It had long since lost its appetite for world domination and had rediscovered its roots as a producer of high-quality medicines. It still ran into controversy from time to time – no enterprise with its pedigree could completely shake off the past. But to all intents and purposes it was now a completely different company – one that had made some significant discoveries, developed a great many new useful drugs and, helped by the German economic miracle of the 1960s and

1970s, evolved into a large modern profit-driven pharmaceutical firm much like any other. With one exception, however: its obsessive determination to recover those parts of its business it had lost to Sterling Products.

For more than forty years Bayer AG (the company shed the Farbenfabriken part of its name in 1972) fought a hugely expensive legal and marketing war with Sterling around the world. It challenged its American rival everywhere it could, launching rival products here, going to the courts there – doing anything, in fact, to get back its name. It even tried to buy it from Sterling in 1964, offering $2.5 million to recover the Bayer brand in territories outside the United States (knowing full well that it had no hope of retrieving it in America). Sterling declined. But if progress was painfully slow, there were still some small victories – in 1976, for instance, Bayer managed to persuade the courts in the Republic of Ireland that Sterling had been misleading consumers into thinking their version of Bayer Aspirin was made in Germany rather than the United States. And gradually, as a result of this extraordinary legal onslaught, the tide turned in Leverkusen's favour. It began to recover the Bayer name almost everywhere in the world.

But not in the United States and Canada. Sterling doggedly refused to retreat one step from the rights it held in its own backyard; engaged in its own life-and-death war with the other American analgesic producers, it was in no mood to concede anything. It responded vigorously and effectively to every challenge Bayer AG made and kept its old adversary at bay. Even when the company was acquired in 1988 by the photographic giant Eastman Kodak (which was trying to expand into the over-the-counter medicines market), there was no let-up in this defence. Eastman knew that Bayer Aspirin was vital to its new subsidiary's success and was in no mood to see it go.

There the matter rested. Bayer AG had to sit idly by, unable to sell any products under its own name in the biggest territory in the world. It did its best to get around the problem by buying Miles Inc. (the makers of Alka-Seltzer) but it was poor substitute for the real thing. In the meantime, of course, aspirin, Leverkusen's proudest creation, was beginning to reveal its potential as a treatment beyond mere analgesia and the Bayer brand in America was in the thick of those developments. The corporate descendants of Carl Duisberg could only gaze on as frustrated bystanders.

And then an opportunity finally appeared. The early 1990s saw the beginning of a massive consolidation in the world pharmaceutical industry as leading drug companies embarked on a merger and acquisition binge. The British-based SmithKline Beecham, itself formed out of several previous mergers, was one of those companies and in August 1994 it struck a deal with Eastman Kodak to buy Sterling Winthrop, as it had become known, for $2.95 billion. Jan Leschly, SmithKline Beecham's ebullient Chief Executive, told the world's press that it was part of a push to make SmithKline a global leader in over-the-counter medicines. He cited Bayer Aspirin as one of the most exciting products in Sterling's US portfolio.

This flag-waving was too much for Leverkusen to bear. Within two weeks Bayer AG had struck a deal with Leschly to buy back its lost North American operations for around $1 billion – a colossal sum and one that SmithKline's shareholders must have been delighted to get. But back in Germany it was seen as a price worth paying. After seventy-five years Bayer AG could once more sell its own products under its own name in the United States. Bayer Aspirin – all of it – was back in Leverkusen's hands and the ghosts of Carl Duisberg and William Weiss were at long last laid to rest.

As the company began its preparations for aspirin's 100th birthday, this hugely satisfying moment was deeply enriched by the knowledge that the little white pill which had launched Bayer to greatness was proving itself, yet again, as the world's most endurable and most surprising medicine. Aspirin's true potential was finally being revealed.

Once aspirin's preventive role on second heart attacks had become widely accepted in the mid-1980s, the next obvious step was to determine whether it could help stop people having a first heart attack.* Logically it should, many cardiologists reasoned, because if it blocked platelet aggregation in one type of heart attack it should in the other. But proving it had seemed a monumental task. Because the ratio of heart attacks in a healthy population was much lower than among those

* Britain's Department of Health gave its official backing to aspirin as a potential treatment for second heart attacks in 1987 when Nicholas Laboratories, makers of Aspro, submitted a licence application for a low-dose aspirin called Platelet 100. Germany, France and Japan had already approved the drug as a heart treatment by then.

people who had already had one, the sample size needed for a clinical trial to demonstrate aspirin's benefits on primary heart attacks was that much larger. However, the proof was even more vital to attain than in secondary coronary cases, because no one wanted otherwise healthy men and women to start swallowing aspirin for no good reason. For the great majority of people, aspirin's side effects in over-the-counter doses are relatively mild and virtually unnoticeable. But for a small minority they do cause problems – bleeding, gastrointestinal distress, and so on – and in a very few cases (either as a result of excessive doses or because someone is particularly allergic to it) these problems can become very serious. And because heart attack had become such a high-profile condition the medical authorities were wary of doing anything to precipitate an aspirin binge, of seemingly encouraging the public to take the drug without medical advice or clear evidence that it might be of benefit. But how to get that evidence when assembling a trial was so difficult?

Richard Peto, Sir Richard Doll and others had set up the British doctors' trial in 1978 to help find this evidence among 5,000 physicians, and had hoped to extend it to the United States. That bid had failed because the US authorities were unconvinced by the self-medication aspects of the study and it was left to an American epidemiologist called Charles Hennekens to have another go. Hennekens had spent some time working in Oxford with Peto and had been convinced by his theories about a very large and very simple trial. In 1981, he had finally managed to convince the National Institutes of Health that it was worth pursuing, and that it need not be prohibitively expensive. The government gave him $3.7 million (somewhat less than the $100 million that a complicated heart disease study might have cost) and Hennekens set about approaching every male doctor over the age of forty in the United States – some 260,000 physicians. In the end he managed to whittle the final sample down to 33,233 subjects (by cutting out those who had any history of heart attack and stroke, or who took high doses of aspirin for rheumatism or arthritis), of whom 22,071 stayed with the trial to the end. The experiment was relatively straightforward. Half got aspirin every other day, half got a placebo every other day. Hennekens awaited the results.

Almost all clinical drug trials have an in-built safety mechanism, an independent committee that monitors the study from start to finish to make sure that it is conducted safely and ethically. One of the most important issues it may have to address is what should happen if the treatment is shown to be clearly beneficial before the trial has run its course. It's a difficult dilemma but it can often come down to a simple question: should the epidemiologist be allowed to continue with the study to make his evidence as conclusive as possible, or should it be halted to avoid giving more placebos to patients who should really be taking the treatment? In the case of something like a heart attack study, this could also clearly be a matter of life and death.

Hennekens' trial had a monitoring committee, comprised of seven experts from the National Institutes of Health. They met every six months to look over his data and to make sure everything was proceeding as planned. At first they had little to do as unhealthy physicians had mostly been excluded from the survey before it began and very few of the rest had died. But by 1988, after the study had been going five years, one very clear trend had become apparent. In total there had been 293 heart attacks to date, of which 189 had been in the control group and 104 had been taking aspirin. The difference between the two groups was in the order of 44 per cent, a figure high enough to suggest that aspirin was having a significant impact. Under those circumstances, was it ethical that patients in the placebo group should be denied the benefits of a treatment that might prevent them dying? There was no real doubt about the answer. The trial was stopped.

When the news broke in January 1989, the media went wild. Television bulletins and newspaper front pages screamed out the news that aspirin gave you a 44 per cent better chance of avoiding a heart attack. Humble aspirin, the little white pill tucked away at the back of your bathroom cabinet, could significantly reduce your chances of succumbing to one of the Western world's biggest killers. Things calmed down a little a few days later when the results of the British doctors' trial were published in the *British Medical Journal*. Although much smaller than its American counterpart, it was still a sizeable sample of over 5,000 and it returned inconclusive results that showed aspirin gave only a small degree of protection. Obviously, one or other of these trials must have been hit by statistical anomalies in some way (they couldn't both be right after all), but once they'd been put through Richard Peto's meta-analysis mill – by then becoming an acceptable way of judging these things – most cardiologists and epidemiologists came to accept the overall conclusion: aspirin might reduce the instance of primary heart attacks by around 33 per cent. It was still a staggering figure.

There was other good news to celebrate. Aspirin, it seemed, was just as effective at preventing another life-threatening condition – stroke. From the moment that scientists realized that aspirin blocked platelet aggregation, doctors began thinking about the consequences across a range of illnesses. Heart failure drew much of the attention because it was such a high-profile condition but clearly any disease that was caused by blood clots was a potential candidate for aspirin treatment. Stroke was an obvious example.

Unlike myocardial infarction, which is often caused by thrombi in the arteries around the heart, stroke is usually a result of blockages in the cerebral arteries.* When a thrombus forms in these arteries it starves the brain of oxygen and can cause permanent and sometimes fatal damage. Around 35 per cent of people who have a major stroke die, the rest can be left blind, paralysed or deprived of memory and mental function. Fortunately, as medicine has come to understand more about this disease, treatment has improved and many first-time stroke victims can recover – particularly if they have had what doctors call a transient ischemic attack (TIA), or in layman's language a temporary interruption in the supply of blood to the brain. However, a TIA is a serious warning sign and people who suffer one or more of these mini-strokes can go on to have a lethal one later. A treatment that could help prevent recurrence of a TIA is clearly of great benefit.

One of the first people to look at aspirin as potential stroke treatment was a doctor from Houston in Texas called William Fields. As early as the mid-1960s he had noticed that people who took aspirin regularly seemed to suffer fewer problems with thrombosis.† When the news broke about aspirin inhibiting platelet stickiness, Fields got together with a New York-based neurologist called William Hass to

* Usually, but not always. Some strokes are caused when a blood vessel bursts in the brain.

† Interestingly, well before aspirin's platelet action was known many ordinary practising doctors around the world had noticed this effect. Although there wasn't yet any scientific evidence to back up what was really just a hunch, some had quietly begun to prescribe aspirin to their patients in much the same way that Lawrence Craven did in California, although he was the only one to write it up as a trial.

develop a trial (in the end it was added on to a more general study that looked at the effects of surgery on stroke patients). With approval from the National Institutes of Health, it got under way in 1971.

Although, like Peter Elwood's first trial, the results were statistically muddy, they did seem to suggest that aspirin might help reduce TIA and major stroke to a significant degree. Regrettably, Fields at first found it difficult to persuade anyone at the National Institutes of Health to take his results seriously. His study (the first part of which was published in 1977) looked at only 178 patients and many neurologists thought this was too small a sample to prove anything. But a few months later another trial from Canada supported his argument. This had been set up to compare aspirin with a second potential platelet-inhibiting drug (a gout medicine called sulfinpyrazone) as a possible stroke treatment. Its results were only a little more conclusive but they did suggest that among male patients with a history of TIA aspirin might reduce the chance of a major stroke by around 30 per cent. Curiously, on women it seemed to make no difference.*

Sterling, whose drug division had provided the Bayer Aspirin and placebo for Field's trial sat up and took notice. It distributed copies of his report, bearing an acknowledgement of its small contribution, to physicians across America. Perhaps inevitably, this came to the attention of the Food and Drug Administration, the agency responsible for monitoring the publicity material pharmaceutical companies sent to doctors about their products. It seemed to the FDA that Sterling was trying to get an unapproved use of aspirin under the wire and it told the drug firm to be careful. Sterling responded by applying to get the label changed – much as it was to do later in response to more high-profile heart attack studies. To the company's great surprise, the FDA's panel of neurology experts gave it their approval.† In 1980, aspirin became an officially recognized stroke treatment. That it didn't attract the same hullabaloo as aspirin's anti-heart-attack role did later says much about the relative status of the two conditions in the minds of the medical profession, the media and the public. But in its way it was as significant a decision. In most

---

* The trials also found sulfinpyrazone to be ineffective.

† Reflecting the ambiguity of the Canadian trial's results for female patients, this approval was initially extended to men only.

Western countries stroke is one of the biggest killers after coronary heart disease.*

The stroke and cardiovascular studies of the 1970s and 1980s, combined with increasing understanding of aspirin's action, opened the floodgates to a huge torrent of research into the drug's potential benefits. Almost every medical specialist, it seemed, began to wonder what acetylsalicylic acid could do for his patients. Many of the clinical trials that followed started small and produced inconclusive results – few things in epidemiological medicine are proved overnight and it often takes several studies to prove things one way or another. But just as many began to excite doctors because of the possibilities that they revealed.

Take senile dementia for instance – a state of mental health categorized by medicine under the umbrella term of cognitive decline. Some people, as they become elderly, start losing their memories and personalities, their understanding of the world around them, even their ability to take care of themselves. There is no real warning of when this creeping degenerative process is about to begin (and because it often happens at a snail's pace it is not so easy to detect at first), but it generally comes about because of two principal causes. The first is Alzheimer's disease, a genetic condition. The second is multi-infarct dementia, the result of several small strokes in the blood vessels of the brain, which over time begin to destroy the cerebral tissue. The consequences of both are equally distressing to sufferers and their relatives.

Because of its effect on stroke, aspirin has been investigated as a possible postponing treatment for both kinds of dementia. The results have been promising. In 1989, John Meyer at the Houston Veterans Administration Medical Center in Texas trialled the drug on a small pilot sample of seventy multi-infarct dementia patients. One half got aspirin, the other didn't. After three years the aspirin takers showed a marked improvement in their cognitive function and several were able to return to work. A British trial some years later tested aspirin and warfarin (an anticoagulant drug) on the cognitive function of 400 men at risk of cardiovascular disease, and found that 'verbal fluency and

---

* It is now thought that in a very few people excessive doses of aspirin can lead to dangerous cerebral bleeding that can cause haemorrhagic stroke – another reason to seek advice before taking the drug.

mental flexibility were significantly better in subjects taking anti-thrombotic medication than in subjects taking placebo. Aspirin may have contributed more than warfarin to any beneficial effect.'

Experiments on aspirin's possible effect on Alzheimer's disease have been equally encouraging. A Baltimore study on the elderly suggested that subjects who used aspirin for two or more years showed a quantifiable decrease in the risk of developing the condition. More dramatically, a more recent and much larger trial, carried out among 5,000 patients over sixty-five within the Puget Sound Health Care System in Seattle, showed that patients who had taken aspirin (or other NSAIDs) for more than two years seemed half as likely to develop Alzheimer's as those who didn't take the pills regularly – provided the drug was taken well before the onset of dementia.

Right at the other end of the age scale, aspirin has also been investigated as a possible treatment for pre-eclampsia. This is a complication of pregnancy, affecting around 8 per cent of mothers, caused to a large degree by tiny thrombi in the placenta. These increase blood pressure, which when combined with an excess of protein in the urine (another consequence of the disease) can cause retarded foetal growth or problems during childbirth. Serious cases can be fatal for mother and child. A big Oxford-based trial, known as the Collaborative Low-dose Aspirin Study in Pregnancy, looked at whether aspirin could prevent the pre-eclampsia thrombi forming but proved inconclusive. A second Oxford study, based on a meta-analysis of thirty smaller trials, has more recently found that anti-platelet drugs (including aspirin) may reduce the risk of pre-eclampsia by around 15 per cent.

Elsewhere, other studies have shown that aspirin may have beneficial effects on a range of less life-threatening conditions, from periodontal gum disease to cataracts and migraines. As is so often the case, there has been debate over how conclusive some of these results are, but the indications are all highly positive.

However, few things have excited the public's attention as much as the revelation that aspirin might be an effective treatment for certain types of cancer. A cure for cancer (the incidence of which grew almost as rapidly as cardiovascular disease in the wealthier nations of the twentieth century) has often been called medicine's search for the Holy Grail. In the last 100 years huge sums of money and millions of man-hours have been invested in this research effort but the results have been slow in coming. Surgery, chemotherapy and radiotherapy techniques have improved to the extent that if some cancers are caught early enough they can be survived, or at least the lives of the victims can be extended. But many are not survivable. In the United States alone 1.3 million new cases occur annually. About half a million Americans die of cancer each year.

Increasingly, therefore, cancer specialists have begun to focus on prevention as well as cure. Clearly, one of the most obvious areas in which this can be effective is in changing people's habits. It is well known that in general cancer shares many causal factors with cardiovascular disease – high-fat diet, sedentary lifestyle, excessive alcohol intake, and of course smoking, the biggest cause of all. Getting people to stop smoking, take more exercise, eat less fatty foods, and so on can all have an impact on their likelihood of developing a carcinoma at some point in their life. But in recent years another approach has been tried. It is known as chemoprevention and involves using various drugs, vitamins, minerals and other chemicals to try to stop tumours forming (or growing if they have already formed). And that's where aspirin comes in.

Back in the mid-1970s, scientists observed that some cancers produce more of a prostaglandin called E2 than surrounding tissue and mucosa. Out of this grew a hypothesis that suggested tumours overproducing E2 in this way might promote their own growth and spread. The more that aspirin's inhibiting action on prostaglandins came to be understood, the more scientists began to wonder whether it might have a role in inhibiting prostaglandin activity related to tumours, focusing attention particularly on the enzyme known as cyclooxygenase 2, which has a role in inflammation and some cancer growth. Early experiments on animals seemed to back up all these theories, and from the mid-1980s, as the science became more sophisticated, they provided some of the impetus for a huge body of clinical aspirin/cancer trials around the world.

Many of those trials have already been published (more seem to appear every few months) and while the results are certainly not yet definitive, and even in a few cases seem to conflict with one another, overall the results have been very encouraging. For instance, a recent review of twenty-seven observational studies on aspirin and colorectal or bowel cancer suggests that the drug (probably as a result of long-term use) may reduce the risk by as much as 50 per cent. Other studies have shown aspirin's potentially preventive effect on cancers of the mouth, throat and oesophagus, prostate cancer (the most common cancer in men), ovarian cancer, breast cancer and lung cancer. As one might expect, the doctors and epidemiologists involved in most of these trials have repeatedly emphasized the need for caution. It is still early days, the results are not totally conclusive, no one is absolutely sure what dosage levels might be appropriate, more time and proper randomized trials are needed, and so on. Furthermore, like those involved in the early cardiovascular studies, they are also concerned lest the public should be provoked into needlessly taking a drug which, though relatively harmless to most people in occasional over-the-counter doses, does have gastric and bleeding side effects in long-term use or high doses.

But the more these studies stack up, the harder it becomes to ignore the conclusion that aspirin's benefits as a cancer treatment might one day be as remarkable as its effects on heart disease. As Professor Chris Paraskeva, an epidemiologist from Bristol University and an expert on the cancer studies, said recently:

> It's difficult not to be optimistic. Aspirin's obviously very useful in bowel cancer for instance – I think enough studies have shown that – and there are good indications it may be effective on these other cancers too. Of course, you have to be cautious about all this because there's still so much work to do and no one wants the general public to misuse aspirin in the belief that it's a cure-all. What is clear though is that it is going to have a major role . . .

Cancer Research UK, one of the largest charities involved in researching the disease, had put it more succinctly. It recently cited aspirin as 'one of the greatest finds in the history of drug discovery'.

The aspirin companies were quick to grasp the commercial possibilities offered by these new uses. The principal catalysts were the stroke and cardiovascular breakthroughs of the 1980s – the FDA decisions on TIAs and secondary heart attacks, and the British and US doctors' trials. From the moment the results of Charles Hennekens' study were first made public and the media went into overdrive, the advertisements began to appear. The idea that an aspirin taken every other day could prevent heart attacks was the kind of material that copy-writers dream of. In America, the FDA tried to dampen everyone's ardour by insisting that it had only given its approval to publicity material about second heart attacks, but when someone pointed out that the public didn't generally make a distinction between first and second heart attacks, it became clear that there was little the authorities could do. Aspirin was big news and there was money to be made. After years of taking a caning at the hands of acetaminophen and ibuprofen, it was pay-back time. Public demand for the drug shot up, and within months, aspirin had once more become the top-selling analgesic in the United States.

Oddly, the bubble didn't last long. By the start of the 1990s brands like Tylenol, Advil and Nurofen began to fight back (trade won in the fierce marketing struggles of the previous twenty years wasn't going to be sacrificed that easily) and aspirin sales began to plateau. They would start to rise again when the longer-term implications of aspirin's renaissance became clearer, but it was an object lesson that medical interest didn't necessarily translate easily into hard cash.

It was also a time of enormous upheaval in the pharmaceutical industry when drug companies became caught up in an acquisition free-for-all. The analgesic business was a counter in this multi-billion-dollar poker game and inevitably many of the aspirin makers featured in this narrative fell victim to it or at least took on other guises to survive. Nicholas Laboratories, makers of Aspro, were bought first by the Sara Lee food conglomerate and then sold on to Roche, the giant Swiss pharmaceutical company. Reckitt and Colman, makers of Disprin, merged with Benckiser to become Reckitt Benckiser. Sterling passed through Eastman Kodak's hands to SmithKline Beecham and then back to Bayer AG. Wellcome, which as Burroughs Wellcome had been one of the first British aspirin producers, joined with Glaxo. The combined group then merged with SmithKline Beecham (of Beecham's Powders fame – another aspirin product) to become GlaxoSmithKline (currently the second largest pharmaceutical company in the world, after Pfizer, the makers of Viagra). Bristol Myers, makers of Advil, Bufferin and Excedrin, merged with the Squibb group. American Home Products reinvented itself as Wyeth. And on and on it went. Old rivals became new partners, old partners became new rivals, and with the waters so muddied it took a while before some semblance of normality returned. In the meantime, of course, the marketing war between analgesic brands continued unabated – and still goes on today. When Johnson and Johnson (unsuccessfully) took Bayer to court in 2002 over claims the aspirin maker had made about Tylenol and another product called Aleve, no one in the industry was that surprised.

As things currently stand in the aspirin business, Bayer is in a strong position to exploit the continued good news about the drug because it currently controls over one third of global aspirin sales. Those have dipped since 1997 in the face of fierce competition from ibuprofen in particular (although the acetaminophen/paracetamol brand Tylenol is still the world's largest), but market analysts believe that between now and 2007, aspirin's gradual repositioning as a 'lifestyle drug' will see its sales grow by about $1\frac{1}{2}$ per cent annually. With Americans alone currently swallowing the equivalent of 80 billion 300 mg tablets every year, that's an awful lot of extra aspirin.

So what does the future hold for this most remarkable of drugs? Well, clearly a great deal more research for a start. Big randomized trials are in the process of being established to try to determine as definitively as possible aspirin's action on cancer and other diseases. Many more will be needed. But it's not such a straightforward process. It has to be remembered that setting up such studies can be extraordinarily costly – even the relatively inexpensive US physicians' cardiovascular trial cost several million dollars and that was twenty years ago. Where is this funding going to come from?

If a trial is aimed at testing out a potentially new pharmaceutical product – one which can be patented and which will deliver several years of high profits if successful – drug companies will invest significant sums of money in finding out if it works because there's a strong chance they'll make their money back. But aspirin is eighty or more years out of patent. It is also extremely inexpensive, retailing at around 1p ($0.015) a tablet. Its producers will always fund a few trials out of altruism but there's no real financial incentive for them to do more because the profit margins on the product are so small and anyone can make it. Why spend tens of millions of dollars on research that your rivals can exploit as easily as you can? Or even, to be more cynical about it, why not wait until academic researchers do the work for you? It's no accident that all the big scientific breakthroughs in aspirin's recent history have come from publicly funded institutions and not from the private sector. For this research to continue – as it must – governments (and by implication the public) will have to be persuaded that the potential benefits to public health, and the potential savings from a reduction in expensive cardiac surgery and oncology wards, are sufficient to justify significant investment.

Money isn't the only issue. Because the new uses of aspirin have received substantial publicity over the past few years, many, if not most, people have an idea of its role in preventing some kinds of cancer and many more know about its role in preventing heart attacks. Yet the next round of trials all need subjects who are prepared to take a placebo (sometimes for several years) and avoid taking a drug that they have already been told has numerous miraculous uses. How many people will be prepared to do that? Indeed, one of the reasons why the big studies into the prevention of primary heart attacks haven't been repeated is because no one has been able to find a way of dealing with the ethical problems of denying patients a drug that might save their life. This is currently less of an issue for the cancer trials because aspirin's action on cancers is less conclusive and the risk is therefore worth taking. But, paradoxically, the more successful they are, the harder it will be to find subjects to take part in future studies. The window for these studies is closing.

Of course, not every trial will be focused on cancer and heart disease. Around 26,000 scientific and medical papers on aspirin have already been published and each year this huge stack of work grows ever larger. At the time of writing there are some 2,000 more research projects in the pipeline. Some of these are taking aspirin into completely different areas of biochemical research and it is perfectly possible that in future aspirin will be shown to have benefits that have not yet been manifest. For instance, some researchers are investigating whether aspirin may have a beneficial effect on the immune system – important for fighting off viruses. No one knows yet where this will lead, but it is not totally implausible to think that one day aspirin might play a role in effective vaccines for diseases like AIDS; it has already been found to inhibit a protein that might allow the virus to multiply.

Research is not restricted to the benefits of aspirin. Others have been looking at ways the drug itself may be remodelled and delivered. One of the most intriguing of these is on the verge of coming to fruition.

It is the brainchild of Professor Kathryn Uhrich, of Rutgers University in Piscataway, New Jersey. In the late 1990s, she was teaching a class of chemistry students and had just asked them to go away and make some acetylsalicylic acid (nowadays a standard undergraduate exercise in laboratory technique) when she had an idea. Professor Uhrich's own specialist research field is in polymers. In simple terms these are long chains of molecules that can form anything from synthetic materials like plastics and nylon to natural materials like rubber, wood, proteins and even DNA. Because the chemistry of polymers can be adjusted – to increase their elasticity, for instance – scientists have long been interested in finding different uses for them. Some years ago, pharmaceutical chemists began developing polymers to use as carrier molecules to deliver drugs – of great use if you want to get a treatment to a particular part of the body where it will be most effective. But until Kathryn Uhrich had her brainwave, no one had ever thought of using them as drugs in their own right. Why not, she thought, make a polymer version of salicylic acid, one of the active parts of aspirin?

Like all good ideas it was a brilliantly simple concept with some fascinating potential. By tinkering with the molecular structure of salicylic acid she was able to come up with something she has called PolyAspirin. The new polymer may have multiple applications. One of its most useful is that it can pass through the stomach, the small intestines and into the bloodstream before it breaks down into salicylic acid, thus leaving the stomach unaffected by gastric side effects. Because the rate at which the polymer degrades can be predetermined, it can also be used as a time-lapsing analgesic – slowly releasing salicylic acid into the body over a long period of time. It is also easy to make it very flexible (due to its similar molecular structure to polyester, the clothing fabric), which means that it can be turned into thread – for use as analgesic surgical sutures, for instance – or, like any other plastic, moulded into shapes, that could be used by orthopaedic surgeons to reduce inflammation around a painful joint or by dentists to fill a painful cavity.

With the help of Rutgers University (to whom she assigned the patent), Professor Uhrich has set up a company to develop the polymer and in 2004 will begin clinical trials. 'I don't see it as a replacement for ordinary aspirin,' she said recently, 'as it will be quite a bit more expensive. But it will allow for other uses of aspirin that haven't previously been possible.'

Others have been thinking along similar lines. A team from Britain's Wolfson Institute of Preventive Medicine are currently designing a 'polypill' that would combine aspirin with an anti-cholesterol drug, three beta-blocker type drugs (that lower blood pressure) and folic acid. Their belief is that together this medicinal cocktail could massively cut heart attack and stroke rates – by up to 80 per cent in people over fifty-five. Costing less than a pound per day per patient, and with minimal side effects (because it would be prescribed only to those with the necessary tolerance), it may have an astonishing impact on the prevention of these diseases in the Western world, and save cash-strapped health services a great deal of money too.

Several thousand years ago a nameless Ancient Egyptian or Sumerian healer inadvertently set off a chain of events that led to one of the most astonishing inventions in history – a drug so incredibly useful and surprising that even now, in the first decade of the twenty-first century, its full potential has yet to be realized.

So how do we best use it?

Aspirin is not a cure-all. It won't stop you breaking your leg or dying from a snakebite or falling prey to depression. It is not even the most potent pain reliever or anti-inflammatory or fever reducer. And although it is generally regarded as one of the safer drugs in existence, it is worth repeating that it *can* have side effects. It can sometimes upset your stomach. Children shouldn't take it at all because of the small but quantifiable risk of Reyes disease. People with a history of high blood pressure, liver or kidney disease, peptic ulcer and other conditions that increase the risk of internal bleeding, should approach it with caution and certainly not without first seeking medical advice.

But set against that are some of the remarkable benefits uncovered in the last twenty years, particularly at the lower doses that cause proportionately fewer gastric and other problems. Quite simply, aspirin can be a life-saver, even in its most old-fashioned form. The American Heart Association says that chewing an aspirin at the first sign of a heart attack can save between 5,000 and 10,000 lives a year. Sir Richard Peto, the Oxford epidemiologist, believes that if everyone at high risk for vascular disease were put on a low daily dose of aspirin about 100,000 fatal and 200,000 non-fatal heart attacks and strokes could be prevented worldwide each year. An NHS review, carried out by researchers in Swansea, South Wales, found recently that the drug dramatically extends life expectancy for everyone aged over fifty: 'Long-term aspirin use could double the chances that individuals have of living into their nineties. A good quality of life would also be expected, given that aspirin may reduce the risk of many diseases associated with ageing.'

Why, then, are not more people taking it? Why was the medical profession initially so slow at prescribing and recommending a drug that can have such an effect? In 1996, it was found that nearly 50 per cent of myocardial infarction patients in the United States did not take aspirin. Two years later, only 45 per cent of myocardial infarction patients at American teaching hospitals were given aspirin, and less than 25 per cent within thirty minutes of admission when the drug is most effective. It was the same story in Europe. As recently as 1994, only 70 per cent of British GPs actually carried aspirin, and less than half the heart patients in Wales were on it – a figure typical of the rest of Britain at the time. Things have massively improved since then, of course, but even now, when there is so much more evidence of its effects on other diseases, epidemiologists are concerned that the message isn't getting through. People are still dying needlessly because they are not taking the drug when they should.

In part this is because doctors do not want people to think that aspirin will mitigate the effects of smoking, unhealthy diet, lack of exercise or any other lifestyle factors that cause or contribute to serious disease. Nor do they want healthy people to start taking the drug unnecessarily. As even Sir Richard Peto said recently, 'It's a question of getting the balance right. It can have a dramatic effect if used properly but if people are not at risk for, let's say, cardiovascular disease, you don't want them to be affected by side effects. The important thing is to get the drug to those people who need it, not to give it to absolutely everyone. That would be counter-productive.'

But it may also be because the idea that aspirin can have such a dramatic effect has still not sunk in with the public. In the old days, when restrictions on advertising were more lax and potential profits were higher, the aspirin makers would have made it their business to let people know what the drug could do for them. Imagine what George Davies, the mercurial marketing genius of Aspro, would have done with the news that 'Aspirin can save your life!' But when did you last see a television advertisement for plain old ordinary aspirin? New brands of ibuprofen and paracetamol are promoted all the time, but never aspirin.

In the absence of a marketing push from the manufacturers, it has to be the responsibility of the medical authorities to keep emphasizing aspirin's benefits to all those adults in high-risk categories – and that might ultimately be almost everyone over the age of forty-five. It should not be difficult to warn those people who should be warned about the side effects, and with a drug that only costs around a penny a tablet it is certainly affordable enough. As the big Western killers, cancer and heart disease, make their way to the developing nations of the Third World, these economic considerations have huge significance too.

One final thought. There will always be people who simply don't feel comfortable with the idea of taking an aspirin every day – even among those whom it would most help. For all its many advantages, aspirin is a synthetically produced chemical product, and there are plenty of people who (rightly or wrongly) prefer herbal or homeopathic treatments to conventional pharmacological medicines. For them, there may be a simpler way to enjoy some of aspirin's benefits, one that goes right back to the origins of this story and the substance from which it sprang – salicylic acid.

Salicylic acid is present in a great many plants – most famously the willow, but in many other trees, grains, fruits and vegetables too. It appears to have many functions, but botanists believe that one of the most important is to initiate a process called apoptosis, whereby a diseased leaf will die and fall off a plant to avoid infecting the rest. One hundred or so years ago, when we bought all our fruit and vegetables at the local greengrocer or market, it was probably the case that among the crops from which those foods came were a sizeable number of diseased or blemished specimens. Nowadays, of course, with the advent of pesticides, sophisticated farming techniques, and our desire to buy perfect goods from the supermarket, we don't see these too often. Indeed, it's a fair bet that unless they are grown organically, all our crops are so 'healthy' that they don't have to produce anything like as much salicylic acid as they once used to. This is not pure speculation. In 2002, researchers in Dumfries in Scotland found that old-fashioned organically grown vegetables had a higher salicylate content than those produced under ordinary modern methods. The differences were small, but they were clearly evident.

So are we missing a trick?

Peter Elwood, although he admits with a smile that he's indulging in 'a little bit of kite flying', believes it is possible that this lack of salicylic acid in our basic foodstuffs might have had a significant impact on our health. He points out, for instance, that cancer and heart disease rates began to rise in the early twentieth century at about the time that farming methods intensified.

> Who knows, maybe nature intended us to eat vegetables with a high level of salicylates, but we've made the mistake of cutting that out. I mean, one of the theories about cancer is that it represents a failure of apoptosis – that a cell with faulty DNA should commit suicide but it doesn't. Maybe aspirin – acetylsalicylic acid – is preventing cancer by enhancing the cell's ability to cure itself. It would be interesting to find out – and who knows where that would lead?

This idea – that nature might have provided us so neatly with both a life-threatening disease and a way of dealing with it, is one that the Reverend Edward Stone would have found intensely appealing; eighteenth-century medicine was underpinned by such notions of checks and balances and he would have seen the beauty of its logic immediately. Perhaps we should be taking more notice of that philosophy today.

What Stone would have made of all the other events that followed his experiments with willow bark is another matter. Probably he would have been appalled and excited in equal measure, although he would surely have been deeply gratified to find out how his own modest theories had contributed to the development of the world's most remarkable remedy. If nothing else, it would have reinforced his belief in the importance of sharing one's wisdom with others – a principle we all have good reason to be thankful for. As he wrote in 1763, 'I have no other motive for publishing this valuable specific, than it may have a fair and full trial . . . and that the world may reap the benefits accruing from it.' The astonishing thing is that those benefits are still emerging.

Aspirin took a long time truly to justify the sobriquet 'wonderdrug', although there cannot really be any argument about this description today. It has a chequered past – on the debit side making some undeserving people a great deal of money and contributing as a result to some of the more unfortunate events in history – but these things happened only because of its enduring appeal and effectiveness as a medicine. Indeed, had it not been for those who ruthlessly exploited its commercial value, it might not have survived long enough to reveal its remarkable therapeutic secrets. On the plus side, of course, countless millions of people have had good reason to be very thankful for its invention and in future years those numbers will grow still further. Aspirin has become a drug for everyman, a treatment so inexpensive and so broadly useful that it is hard to imagine what we would do without it. There are few products of human ingenuity about which that can be said.

NOTES AND SOURCES

BOOK ONE

1: IF YOU EXAMINE A MAN

p.5 Edwin Smith is something of an enigma. There are very few references to him in the literature of Egyptology and I have therefore had to piece this portrait together from scattered scraps of reminiscence and anecdote that I've found dotted around here and there.

p.5 'But the brothers Ahmed and Mohammed . . .' Both the Abd er-Rasul family's status as Luxor's most successful grave-robbers *and* the fact that they were in the service of Mustapha Agha are elaborated on in John Wilson's *Signs and Wonders Upon Pharaoh: A History of American Egyptology* (University of Chicago Press, 1964). They came into particular prominence in the 1880s when they disclosed the location of a hiding place of royal mummies, and Mohammed was rewarded with a government job. While it is impossible now to confirm that it was the brothers who procured the papyri for Smith, it seems highly likely, as they were Agha's chief suppliers and Smith had bought numerous articles from them before.

p.6 'found between the feet of a mummy . . .' John F. Nunn, *Ancient Egyptian Medicine* (London: British Museum Press, 1996).

'if you examine a man' The opening words of the Ebers Papyrus as translated into English. B. Ebell, *The Papyrus Ebers: The Greatest Egyptian Medical Document* (Copenhagen: Levin and Munksgaard, 1937).

'Edwin Smith was born . . .' These biographical details are drawn variously from Wilson, *Signs and Wonders*, which contains a brief character sketch; W. R. Dawson and E. P. Uphill, *Who Was Who in Egyptology* (London: Egyptian Exploration Society, 1993) and, most significantly, from J. H. Breasted, *The Edwin Smith Surgical Papyrus* (University of Chicago Press, 1930) – see General Introduction. That Smith came from a wealthy family can in part be deduced from a portrait of him painted in 1847 by Francesco Anelli, a New York society artist, when Smith was about twenty-five years of age. It is currently held by the New York Historical Society. See *Catalogue of American Portraits* (New Haven: Yale University Press, 1974). An oblique reference to the events that caused Smith to leave America is contained in *The Letters of Mrs Henry Adams 1865–1883*, edited by Ward Thorn (Boston, Mass.: Little Brown, 1936), p. 75. The exact nature of that scandal is unknown but I would hazard a guess that it was of a sexual nature.

p.7 'becoming a money-lender . . .' Dawson and Uphill, *Who Was Who in Egyptology.* 'Lucy Duff Gordon . . .' The friendship between Smith and Lady Duff Gordon is cited in 'The Place of the New York Historical Society in the growth of American interest in Egyptology' by Caroline Ransom Williams, *The New York Historical Society Quarterly Bulletin*, April 1920. See also Lucy Duff Gordon, *Letters from Egypt* (London: R. Brimley Johnson, 1902).

Examples of Smith's brief correspondence with Charles Wycliffe Goodwin can be found among Goodwin's papers in the British Library (Add. MSS 31268–98).

'his landlord, Mustapha Agha . . .' Wilson, *Signs and Wonders*. A further insight into the sometimes turbulent relationship between Smith and Agha can be gained from an anecdote in A. L. Adams's *Notes of a Naturalist* (Edinburgh: Edmonston and Douglas, 1870), which describes an attempts by Smith to beguile two visiting British scholars into buying some fake antiquities. The plan was subsequently exposed by Agha who then, no doubt, sold the pair some forged artefacts of his own. Another example of the pair's inventiveness concerns a visit by the Prince of Wales to Luxor. Smith and Agha laid on an archaeological demonstration for him and 'found' thirty mummies and various rare artefacts in an underground tomb. These they brought to the surface to the astonishment and admiration of the royal party. See S. Birch, *Transactions of the Royal Society of Literature* (London, 1870).

Smith has also been accused of playing practical jokes on his fellow Egyptologists by carving hieroglyphic graffiti in the Luxor Temple – symbols that confused successive generations of experts who believed them to be authentic but couldn't agree on what they meant. See William J. Murnane, *The Princess Who Never Was: A Tale of Scholarly Agonizing, Piracy and Revenge* (Chicago: The Oriental Institute News and Notes, 1984).

p.8 'on 20 January 1862 . . .' Breasted, *The Edwin Smith Surgical Papyrus*. Footnote: ibid.

'it was as though . . .' ibid.

p.9 'At 110 pages . . .' Nunn, *Ancient Egyptian Medicine.* See also Ebell, *The Papyrus Ebers*.

'*wekhudu* . . .' – sometimes written as 'whdw'. See the *Oxford Encyclopaedia of Ancient Egypt: Medicine*, edited by Donald B. Bedford (Oxford University Press, 2001).

'extensive pharmacopoeia . . .' See Roy Porter, *The Greatest Benefit to Mankind: A Medical History of Humanity from Antiquity to the Present* (London: HarperCollins, 1997).

p.10 'Less than 10 per cent . . .' From information provided by the Centre for Economic Botany, Royal Botanical Gardens, Kew, London.

'The true origins of willow's use as a medicine . . .' The notion that willow (and other plants) were used as medicine by prehistoric peoples and that such use developed as a consequence of observation, ritual, trial and error is a distillation of views found in several standard texts on the history of medicine.

'over 300 known variants of the willow species . . .' From information provided by the Centre for Economic Botany, Royal Botanical Gardens, Kew, London.

p.11 'The very earliest known written reference . . .' See H. Avalos, *Illness and Health Care in the Ancient Near East* (Scholars Press, 1995); J. Bottero, *Everyday Life in Ancient Mesopotamia* (Baltimore: Johns Hopkins University Press, 2001).

'simple trade links . . .' Peter Jay, *The Road to Riches* (London: Weidenfeld & Nicholson, 2000).

Footnote: for example, see I. Raskin, 'Role of salicylic acid in plants', *Annual Review of Plant Physiology, Plant Molecular Biology*, 1992, and W. S. Pierpoint, 'The natural history of salicylic acid', *Interdisciplinary Science Reviews*, 1997.

p.12 Medicinal uses of willow in Ebers Papyrus: gleaned from Ebell, *The Papyrus Ebers*.

p.13 'They didn't yet use opiates . . .' Nunn, *Ancient Egyptian Medicine*.

'Edwin Smith was told . . .' Breasted, *The Edwin Smith Surgical Papyrus*.

'The medical profession, held in high esteem . . .' P. Ghalioungui, *The Physicians of Pharaonic Egypt* (Mainz Verlag Phillip von Zabern: 1983).

p.14 Hippocratic medicine: there is a vast body of research on this subject but the sources I found most useful were G. E. R. Lloyd (ed.), *Hippocratic Writings* (Harmondsworth: Penguin, 1978); James N. Longrigg, *Greek Rational Medicine* (London: Routledge, 1991) and Porter, *The Greatest Benefit to Mankind*.

p.16 Letter from Edwin Smith to Charles Goodwin: Goodwin's papers at British Library (Add. MSS 31268–98).

Advertisement for papyrus quoted by Breasted.

'Ebers . . . tarnished his reputation . . .' In letters to academic journals Professor Ebers vehemently defended his claim that it was he rather than Edwin Smith who had discovered the papyrus, e.g. a note to *Zeitschrift für Aegyptische Sprache und Alterthumskunde* (vol. 11, 1873), but there is now no doubt that he was falsely claiming credit that he was not due.

On his death in Naples . . .' The family scandal that had forced Edwin Smith to leave the United States appears to have blown over by the time he died – sufficiently, at least, to allow him to be reconciled with his daughter, Leonora, who was with him at the end.

2: THE BARK OF AN ENGLISH TREE

p.17 'My Lord . . .' Edward Stone, 'An account of the success of the bark of the willow in the cure of agues,' *Philosophical Transactions*, Royal Society of London, 1763.

'There is a bark of an English tree . . .' ibid.

p.18 Much of the biographical detail about Edward Stone contained in these pages comes from material generously shared with the author by the Reverend Ralph Mann who has spent many years gathering data on Stone's life. His collection of papers includes various administrative documents from Buckinghamshire, Essex and Oxford Country Record Offices and Wadham College, Oxford, all of which I have drawn on here. Mann's own assessment can be found in an article for the *Dictionary of National Biography*, 1993. Other useful sources are W. S. Pierpoint's paper, *Edward Stone (1702–1768) and Edmund Stone (1700–1768): confused identities resolved* (Notes Rec. Royal Society of London 51(2), 211–17, 1997), and P. Fairley, *The Conquest of Pain* (London: Michael Joseph, 1978). The descriptions of Chipping Norton are based on my own visits to the town (the centre is still pretty much as it was in Stone's time) and its small but helpful museum, which contains contemporary maps and prints.

Footnote: 'During a famous election . . .' R. J. Robson, *The Oxfordshire Election of 1754* (Oxford University Press, 1949).

pp.18–19 'at their seat of Bruern . . .' the original house burnt down later in the eighteenth century and the replacement, which still stands, is now a private school.

p.20 Footnote: See Anon, A Faithful Narrative of the Proceedings in a Late Affair Between the Rev. Mr John Swinton, and Mr George Baker, Both of Wadham College, Oxford. to Which is Prefix'd, a Particular Account of the Proceedings Against Robert Thistlethwayte. for a Sodomitical {sic} Attempt Upon Mr W. French, Commoner of the Same College. (London, 1739).

'Two handsome parlours . . .' Document in Ralph Mann's possession.

Eighteenth-century medicine. The most useful reference sources I have found for this subject are: A. Cunningham and R. French (eds), *The Medical Enlightenment of the Eighteenth Century* (Cambridge University Press, 1990); Lester S. King, *The Medical World of the Eighteenth Century* (University of Chicago Press, 1958); D. Porter and R. Porter, *Patients Progress: Doctors and Doctoring in Eighteenth Century England* (Cambridge: Polity Press, 1989); R. Porter, *Blood and Guts* (London: Allen Lane, Penguin Press, 2002); and Eric Jameson, *The Natural History of Quackery* (London: Michael Joseph, 1961).

p.22 'Cur'd yesterday of my disease . . .' Matthew Prior, 1664–1721, satirist and poet, remembered for his witty epigrams.

'well past the average life expectancy . . .' See T. McKeown, *The Modern Rise of Population* (New York: Academic Press, 1976).

p.23 Geoffrey Chaucer, 1343–1400. Quote from *The Canterbury Tales* (London: Folio Society, 1986).

p.24 'Agues are occasioned . . .' William Buchan, *Domestic Medicine or the Family Physician* (Edinburgh: Balfour, Auld and Smellie, 1769).

The *Anopheles* mosquito and malaria in England: see M. J. Dobson, '"Marsh Fever": the geography of malaria in England', *Journal of Historical Geography*, 1980, and P. Reiter, 'From Shakespeare to Defoe: Malaria in England in the Little Ice Age', *Emerging Infectious Diseases*, vol. 6, 2000.

p.25 For more on the discovery of cinchona bark see M. L. Duran-Reynals, *The Fever Bark Tree: The Pageant of Quinine* (New York: Doubleday, 1946) and R. Klein, 'The fever bark tree', *Natural History*, vol. 85, 1976.

p.26 'Robert Talbor . . .' ibid.

p.27 Paracelsus: see Roy Porter, *The Greatest Benefit to Mankind: A Medical History of Humanity from Antiquity to the Present* (London: HarperCollins, 1997), and Allen Debus, *The Chemical Philosophy* (New York: Science History Publications, 1977).

p.28 'What Kench had to say . . .' P. Fairley, *The Conquest of Pain* (London: Michael Joseph, 1978).

'It seemed probable . . .' Stone, 'An account of the success of the bark of the willow in the cure of agues'.

Nicholas Culpeper, 1616–54. Author of *The English Physician: or an atrol-ogo-physical discourse of the vulgar herbs of this nation* (London: Peter Cole, 1652).

p.29 'It was not long before . . .' Stone, 'An account of the success of the bark of the willow in the cure of agues'.

p.30 'I have continued . . .' ibid.

'The business of the society . . .' From *The Charter of the Royal Society of London for the Promotion of Natural Knowledge 1663*. See also Sir Henry Lyons, *The Royal Society 1660–1940* (Cambridge University Press, 1994); M. B. Hall, *Promoting Experimental Learning – experiment and the Royal Society 1660–1727* (Cambridge University Press, 1991) and E. N. da C. Andrade, *A Brief History of the Royal Society* (London: Royal Society, 1960).

p.32 'a good man, but slow . . .' R. Mann in *Dictionary of National Biography – Missing Persons* (Oxford University Press, 1993).

'I have no other motive . . .' Stone, 'An account of the success of the bark of the willow in the cure of agues'.

Earl of Macclesfield's death; see Lyons, *The Royal Society*.

p.33 'A James Burrows Esq. sat in the chair . . .' *Royal Society Journal Book*, 2 June 1763.

'the letter, dismissed as unoriginal, was never published . . .' ibid., 19 November 1767.

p.34 Stone died intestate at Chipping Norton on 26 November 1768 and was buried at Horsenden, Buckinghamshire.

'the singular efficacy . . .' S. Jones, *Observations on the Bark of a particular species of willow, showing its superiority to the Peruvian and its singular efficacy in the cure of agues etc.* (London: Johnson, 1792). See also J. Mann *Murder, Magic and Medicine* (Oxford University Press, 1992). White quote in *Bath Register*, November 1798.

'But quinine attacks . . .' The alkaloids in cinchona bark interfere with the reproduction of the plasmodium parasite.

3: THE PUZZLE TAKES SHAPE

p.37 Johann Christian Reil: a scientific philosopher who grew out of the Romantic movement. Perhaps his most famous contribution to medicine was his foundation of the *Journal for Psychological Therapy* in 1805, which later became a cornerstone of German psychiatry.

For the continuing demand for cinchona bark and the problems of obtaining it, see M. L. Duran-Reynals *The Fever Bark Tree: The Pageant of Quinine* (New York: Doubleday, 1946).

p.38 'the experimental science . . .' See W. Brock, 'The Biochemical Tradition' in *Companion Encyclopaedia of the History of Medicine* (London: Routledge, 1993); Erwin H. Ackerknecht, *Medicine at the Paris Hospital, 1794–1848* (Baltimore: Johns Hopkins University Press, 1967); R. Porter, *The Greatest Benefit to Mankind: A Medical History of Humanity from Antiquity to the Present* (London: HarperCollins, 1997); and Hans Schadewaldt and Rosemary Alstaedter, *History of Pharmacological Research at Bayer* (Bayer, 1991).

p.39 'Johann Pagenstecher was already . . .' See *Revue de Pharmacologie* (Paris, 1926) and P. Fairley, *The Conquest of Pain* (London: Michael Joseph, 1978).

p.40 Lowig: ibid.

p.41 The role of coal in the Industrial Revolution: T. Warner, *Landmarks in Industrial History* (London: Blackie and Son, 1909).

Footnote: B. Taylor, *William Murdoch: New Lamps for Old* (London: Macmillan, 1952).

p.42 Friedlieb Ferdinand Runge: see Schadewaldt and Alstaedter, *History of Pharmacological Research at Bayer*; W. H. Brock, *The Fontana History of Chemistry* (London: Fontana Press, 1992).

p.43 William Perkin: see S. Garfield, *Mauve: How One Man Invented a Colour that Changed the World* (New York: Norton, 2001); A. S. Travis, *The Rainbow Makers: the origins of the synthetic dyestuffs industry in Western Europe* (Bethlehem: Lehigh University Press, 1983) and the website of Imperial College, London, www.ch.ic.ac.uk

p.44 'For almost as long . . .' See Travis, *The Rainbow Makers*.

p.46 'a French scientist called Charles Gerhardt . . .' P. Bachoffner, 'Two pharmacists in the beginning of aspirin', *Revue d'Histoire Pharmaceutique*, Paris, 1996, and Klaus Florey, *Analytical Profiles of Drug Substances*, vol. 8 (London: Academic Press, 1979).

p.48 Daniel Defoe: *A Tour through the Whole Island of Great Britain 1724–6* (London, 1727).

Industrial development of Dundee: Michael Lynch (ed.), *Oxford Companion to Scottish History* (Oxford University Press, 2001); and for a contrasting view: Louise Miskell, Christopher Whatley and Bob Harris, *Victorian Dundee* (Tuckwell Press, 2001).

p.49 Consequences of bad housing and poverty: see *Report of the Sanitary Condition of the Labouring Population of Scotland of 1842* (Public Records Office, Kew, London).

Biographical details on Maclagan: William K. Stewart and L. W. Fleming, 'Perthshire pioneer of anti-inflammatory agents', *Scottish Medical Journal*, 32:141/146, 1987.

'The city was in the grip . . .' H. Gibson, *Dundee Royal Infirmary 1798–1938* (Dundee: Kidd, 1948).

p.50 'On many occasions when visiting . . .' T. J. Maclagan, 'Typhus statistics of the Dundee Royal Infirmary', *Edinburgh Medical Journal*, vol. xiii, 1867–8.

'caught only a mild form of the disease . . .' T. J. Maclagan 'On enteric fever in Dundee and neighbourhood', ibid.

'lowest patient-mortality rates in Scotland . . .' Maclagan, 'Typhus statistics of the Dundee Royal Infirmary'.

p.51 'Maclagan's own strange pet belief . . .' *Lancet*, 1903.

p.52 'The DRI was an impressive institution . . .' Gibson, *Dundee Royal Infirmary 1798–1938*.

'Nature seeming to produce the remedy . . .' T. J. Maclagan, 'The treatment of acute rheumatism by salicin', *Lancet*, 4 March 1876.

p.53 'I had at that time . . .' ibid.

'two shillings an ounce . . .' *British Pharmacopoeia*, 1876.

'the strictest views . . .' A remark from Sir Frederick Treeves, Maclagan's personal doctor, *British Medical Journal*, March 1903.

p.54 'quite apart from its antipyretic properties . . .' Maclagan, 'The treatment of acute rheumatism by salicin'.

'the price of salicin . . .' *British Pharmacopoeia*, 1877.

'Solomon Stricker . . .' See *Ueber die Resultate der Behandlung der Polyartritis rheumatica mit Salicylsäure* (Berl. Klin. Wochenschr., 1876).

'Ludwig Reiss . . .' See *Nachtrag zur innerlichen Anwendung der Salicylsäure ins besondere bei dem acuten Gelenkrheumatismus*, ibid.

'Germain See . . .' See *Bulletin Academique de Médecine*, vol. 6, Paris, 1877.

'Dr Ensor of the Cape of Good Hope . . .' *Lancet*, 1876.

p.55 'Maclagan moved south to London . . .' *British Medical Journal*, March 1903; G. Sharp, *Pharmaceutical Journal*, vol. 94, 1915.

'Decoction of willow bark . . .' *Lancet*, March 1903.

4: THE BIRTH OF A WONDER DRUG

p.56 London International Exhibition: details of exhibits from *The Times*, 20 June 1862, and William Perkin from S. Garfield, *Mauve: How One Man Invented a Colour that Changed the World* (New York: Norton, 2001).

pp.57–8 'it was a French chemist . . .' Meredith Dorner, *Early Dye History and the Introduction of Synthetic Dyes before the 1870s* www.smith.edu./hsc/silk/papers/dorner, and A. S. Travis, *The Rainbow Makers: the origins of the synthetic dyestuffs industry in Western Europe* (Bethlehem: Lehigh University Press, 1983).

p.58 'They grasped the opportunity with both hands . . .' J. Beer, *The Emergence of the German Dye Industry* (Illinois Studies in Social Sciences, University of Illinois Press, 1959).

p.58 'Friedrich Bayer and Johann Friedrich Weskott . . .' E. Verg, G. Plumpe and H. Schultheis, *Milestones* (Bayer AG, 1988).

p.59 Biographical details on Carl Duisberg's early life, education and his appointment at Farbenfabriken Bayer from C. Duisberg, *Meine Lebenserinnerungen* (P. Reclam. Jour., Leipzig 1933); Verg et al., *Milestones;* H. Armstrong, 'Chemical Industry and Carl Duisberg', *Nature*, 22 June 1935; and H. Flechtner, *Carl Duisberg: vom Chemiker zum Wirtschaftsführer* (Econ. Düsseldorf, 1959).

pp.62–3 Discovery of antipyrine and Antifebrine: Verg et al., *Milestones*, and B. Issekutz, *Die Geschichte der Arzneimittelforschung* (Budapest, 1971).

p.63 'when a drug appeared with a nice simple name . . .' J. McTavish, 'What's in a name? Aspirin and the American Medical Association', *Bulletin of Historical Medicine*, vol. 61, 1987.

p.64 Development of Phenacetin: Verg et al., *Milestones*.

p.65 Development of Sulfonal: Hans Schadewaldt and Rosemary Alstaedter, *History of Pharmacological Research at Bayer* (Bayer, 1991).

'Chemists had had to find space . . .' ibid.

p.66 'Find new ways of presenting . . .' ibid.

'It was just as well then . . .' ibid.

p.67 Biographical details on Felix Hoffman: Verg et al., *Milestones.*

'*Etablissementserfindung* . . .' Schadewaldt and Alstaedter, *History of Pharmacological Research at Bayer*.

The appointment of Arthur Eichengrün: ibid, and from information provided to the author by Ernst Eichengrün.

p.68 Biographical details on Heinrich Dreser: ibid. and *Sunday Times*, 13 September 1998.

'If a much promulgated company legend . . .' This version of events can be found in the official Bayer history – Verg et al., *Milestones*, and on the company's website see http://www.bayeraspirin.com 'Who Discovered Aspirin?'

'In 1897 . . .' Although there is considerable controversy over this issue, in my opinion the most reliable first-hand account of the discovery of aspirin was written by Arthur Eichengrün in 1949. See A. Eichengrün, '50 Jahre Aspirin', *Pharmazie*, 1949. However, he also wrote an earlier, simpler account: see A. Eichengrün, *Pharmaceutisch-wissenschafliche Abteiling*. In: *Geschichte und Entwicklung der Farbenfabriken vorm Friedr Bayer & Co, Elberfeld, in den ersten 50 Jahren* (Munich: Meisenbach-Riffrath, 1918). As can be seen in Chapter 9 of this book, this 1918 paper later helped fuel the debate over Eichengrün's contribution to the discovery of aspirin because it made so little of his own role in the affair. But it must be remembered that the drug's paternity wasn't in question at the time it was being written and there was therefore no reason for Eichengrün to stake a claim to his share of the credit. (It wasn't Bayer practice to do so, and in any case, he assumed, quite naturally, that he already had it.) Only later, when it appeared that credit had been denied him, did he respond with a more detailed paper asserting his rights. As a consequence the account in these pages is based on Eichengrün's 1949 version, which has to be considered the more accurate of the two.

p.70 'When salicylic acid . . .' Felix Hoffman Laboratory Journal, 10 August 1897 (Bayer Leverkusen Archives).

p.71 'So far so good . . .' Eichengrün, '50 Jahre Aspirin'.

Heinrich Dreser, Felix Hoffman and the discovery of heroin: Schadewaldt and Alstaedter *History of Pharmacological Research at Bayer*; *Sunday Times*, 13 September 1998; and *Bulletin of Narcotics* (April 1953).

p.72 'Dreser was telling . . .' See Heinrich Dreser speech, 19 September 1898, entitled *Pharmakologisches über einige Morphinderivate* (Bayer Leverkusen archives).

'Eichengrün sent off small . . .' Eichengrün, '50 Jahre Aspirin'.

p.73 'This is the usual Berlin boasting . . .' ibid.

'On 23 January 1899, a memo . . .' (Bayer Leverkusen archives). It shows that the name of aspirin was not derived, as has sometimes been supposed, from that of St Aspren, a Neapolitan saint with a gift for healing!

p.73 'Later that year, Dreser did his duty . . .' *Pharmakologisches über Aspirin-Acetylsalicylsäure* (Archiv für die Gesammte Physiologie, 1899).

p.74 'If Dreser was that disgruntled . . .' Eichengrün, '50 Jahre Aspirin'.

H. O. J. Collier, 'The story of aspirin', *Discoveries in Pharmacology*, vol. 2: *Haemodynamics, Hormones and Inflammation*, edited by Parnham and Bruinvels (Elsevier, 1984).

BOOK TWO

5: PATENTS, PATIENTS, AND SELL, SELL, SELL!

p.77 'In the late summer of 1899 . . .' Promotional leaflets *Aspirin*, 1899 (Bayer Leverkusen archives).

p.78 'The first doctor to get hold . . .' K. Witthauer, Ther. Mh. vol 13, 1899.
'Another pre-launch trialist . . .' J. Wohlgemut, ibid.
'Within three years . . .' F. Wohr (Medical Bulletin Phil 1902).
'To Bayer's delight . . .' C. Mann and M. Plummer, *The Aspirin Wars, Money, Medicine and 100 Years of Rampant Competition* (New York: Knopf, 1991).

p.79 'The British patent was filed . . .' Bayer & Co Letters Patent 27,088 (1898).
'The same justification . . .' US Patent No. 644,077. Granted on 27 February 1900, the patent was for seventeen years.

p.80 Lydia Pinkham: E. Applegate, *Personalities and Product: A Historical Perspective on Advertising in America* (Greenwood Press, 1998).

p.81 'she would be ironically celebrated . . .' By the pop band The Scaffold. In 1968 this song spent twenty-eight weeks in the UK charts.

p.82 'The American Medical Association . . .' J. McTavish, 'What's in a name? Aspirin and the American Medical Association', *Bulletin of Historical Medicine*, vol. 61, 1987.
'But when the German chemical companies . . .' J. McTavish, *Aspirin in Germany: The Pharmaceutical Industry and the Pharmaceutical Profession* (Pharmacy in History, 1987).

p.83 'The firm's American business . . .' E. Verg, G. Plumpe and H. Schultheis, *Milestones* (Bayer AG, 1988).

p.84 'Duisberg had bombarded . . .' H. Flechtner, *Carl Duisberg: vom Chemiker zum Wirtschaftsführer* (Econ. Düsseldorf, 1959).
'He was determined . . .' ibid.
Purchase of Rensselaer: Verg et al., *Milestones*.
'Shortly before eleven . . .' Details of this case are to be found in *Farbenfabriken vormals Friedrich Bayer & Co v. Chemische Fabrik Von Heyden* (Reports of Patent, Design and Trade Mark Cases, 1905, 22: 501–18).

p.86 'Chemische Fabrik Von Heyden had been . . .' McTavish, *Aspirin in Germany*.

pp.86–7 Justice Joyce in court: *The Times*, May 1905.

p.87 'It was "erroneous and misleading . . ."' *Farbenfabriken vormals Friedrich Bayer & Co v. Chemische Fabrik Von Heyden*.

p.88 'Bayer's loss of its British aspirin patent . . .' *Lancet*, October 1905.

p.89 'Big foreign syndicates . . .' Lloyd George quoted in J. Borkin, *The Crime and Punishment of I.G. Farben* (New York: Free Press, 1978).
'The more people in high places hinted at foreign conspiracies . . .' The way this growing tension was affected by popular culture is brilliantly explained in Niall Ferguson's *The Pity of War* (London: Allen Lane, Penguin Press, 1998).

p.90 'for some time there had been complaints . . .' McTavish, *Aspirin in Germany*.
'Bayer had filed a similar infringement suit in Chicago . . .' See *Farbenfabriken of Elberfeld Co. v Kuehmsted* (171 Federal Reporter, 1909).
'Aspirin has in the decade since its . . .' Quoted in J. McTavish, *The German Pharmaceutical Industry 1880–1920: A Case Study of Aspirin* (Master's thesis, University of Minnesota, 1986).

p.92 'Its editor, Norman Hapgood . . .' James Harvey Young, *The Toadstool Millionaires: A Social History of Patent Medicines in America before Federal Regulation* (Princeton University Press, 1961). As Young points out, Hapgood and Adams weren't the only campaigning journalists in this field. Another was Mark Sullivan of the *Ladies' Home Journal*. One of his most memorable scoops was finding and photographing Lydia Pinkham's grave in Pine Grove Cemetery in Lynn, Massachusetts – proving once and for all that she couldn't possibly be responding to all those letters asking her for advice.

p.93 'Gullible America will spend this year . . .' Samuel Hopkins Adams, 'The Great American Fraud', *Collier's Magazine*, 7 October 1905.
'Born in 1844 on a small farm . . .' H. W. Wiley, *Harvey W. Wiley – An Autobiography* (Indianapolis: Bobbs-Merrill Co., 1930).
'A large, bull-like figure . . .' E. Bjorkman, *Our Debt to Dr Wiley* (World's Work, 1910).
'The more scandals he exposed . . .' Young, *The Toadstool Millionaires*.

p.94 'The combined public outcry . . .' ibid.
'Wiley instituted proceedings . . .' See *US v Harper 1908, Washington Evening Star*, 16–19 March 1908.

p.95 'The drug had proved so popular . . .' Mann and Plummer, *The Aspirin Wars*. This figure also appears in Verg et al., *Milestones*.

p.96 'Its solution was to start making the tablets itself . . .' Verg et al., *Milestones* and McTavish 'What's in a name?'
'But not everyone was fooled . . .' *Druggists' Circular Chemists' Gazette* vol. 63, 1919. However, the confusion surrounding the various chemical designations of aspirin was so great that some doctors even wrote to their professional journals asking if ASA and aspirin were the same thing. See A. A. Ransom, *Journal of the American Medical Association*, vol. 46, 1906.

6: THE CHEMISTS' WAR

p.97 'I am pleased to see . . .' Letter to the *Lancet*, October 1914. Perceptive readers will have noticed that the signatory of this letter shares the unusual surname, Treeves, with the close friend and personal physician of Dr T. J. Maclagan (see notes for page 53). Regrettably, I have been unable to find out if they were related, which is a shame because this would have had a pleasing symmetry.

'The process began in Britain . . .' See Niall Ferguson, *The Pity of War* (London: Allen Lane, Penguin Press, 1998) and Barbara W. Tuchman, *The Guns of August* (London: Constable, 1962).

pp.98–9 'most of the non-Bayer ASA . . .' Imports of German acetylsalicylic acid into pre-war Britain accounted for over 85 per cent of the market (*Reports*, Pharmaceutical Society, 1912).

p.99 'The best-selling locally manufactured brand . . .' Burroughs Wellcome's Xaxa brand was given its trademark in 1906, one of the first non-German companies to trademark its ASA in this way. J. McTavish, *Aspirin in Germany: The Pharmaceutical Industry and the Pharmaceutical Profession* (Pharmacy in History, 1987) (notes).

'as the British Pharmaceutical Society reminded its members . . .' Mann and Plummer, *The Aspirin Wars, Money, Medicine and 100 Years of Rampant Competition* (New York: Knopf, 1991).

p.100 'The Aspirin Trade Mark . . .' *Lancet*, 5 February 1915.

'One journal boasted that . . .' *The Prescriber*, March 1915.

'A British Army doctor . . .' Letter from Lt E. Bridge, RAMC, to Dr T. N. Ripley, Glasgow, September 1915. Copy in author's possession.

'At least the army was getting some . . .' *The Prescriber*, May 1815.

'Even though Bayer's British subsidiary . . .' The Board of Trade ordered Bayer's UK subsidiary, Bayer Company Ltd, to cease trading in 1916.

p.101 'George Nicholas patted out . . .' R. Grenville-Smith and A. Barrie, *Aspro – how a family business grew up* (Nicholas International Ltd, 1976); B. Morgan, *Apothecary's Venture: The Scientific Quest of the International Nicholas Organisation* (Nicholas Kiwi, 1959); P. Fairley. *The Conquest of Pain* (London: Michael Joseph, 1978); and Stephen Nicholas, information provided to the author.

pp.103–4 'Australian Aspirin is Granted Licence . . .' *Melbourne Herald*, 17 September 1915.

p.104 'With almost no money . . .' Grenville-Smith and Barrie, *Aspro*.

p.106 'Should Aspirin Be Re-named?', *Australasian Journal of Pharmacy*, 21 May 1917.

'Years later the company would . . .' Morgan, *Apothecary's Venture*.

'George Davies . . .' Grenville-Smith and Barrie, *Aspro*.

p.107 'It was then, Davies recalled later . . .' *Chemist and Druggist Diary*, 1927.

p.108 'This infuriated not only . . .' Tuchman, *The Guns of August*.

'Chemicals, particularly coal-tar products . . .' Williams Haynes, *American Chemical Industry – The World War I Period: 1912–1922*, vol. 2 (New York: D. Van Nostrand Company, Inc., 1945).

p.109 'German sympathizers in the United States . . .' Howard Watson Ambruster, *Treason's Peace: German Dyes and American Dupes* (New York: The Beecham Press, 1947).

'Ownership of Bayer's US assets and trademarks . . .' *New York Times*, 19

November 1915, and *Records of the Office of Alien Property (131)*, US National Archives, Washington, DC.

pp.109–10 Phenol as vital element in manufacture of explosives: Haynes, *American Chemical Industry*.

p.110 'Schweitzer had emigrated . . .' Mann and Plummer, *The Aspirin Wars*.

Footnote: See S. Garfield, *Mauve: How One Man Invented a Colour that Changed the World* (New York: Norton, 2001).

p.111 'When the conflict broke out . . .' Ambruster, *Treason's Peace*.

'Schweitzer robustly defended the act . . .' *New York Times*, 1 November 1917.

'In no other field . . .' Schweitzer, quoted in *Aims and Purposes of The Chemical Foundation Inc and the Reasons for its Organisation. As told by A. Mitchell Palmer, United States Attorney General and Former Alien Property Custodian in his report to Congress, and by Francis P. Garvan, Alien Property Custodian, in an address to the National Cotton Manufacturers Association* (New York: De Vinne Press, 1919).

'Recruited by the German Secret Service . . .' Dyestuffs Committee on Ways and Means, US House of Representatives, 66th Congress. Hearings held on 18 June 1919 (US Library of Congress). See also J. P. Jones, *The German Secret Service in America 1914–18* (Toronto: William Briggs, 1918).

'The German Ambassador at the time . . .' ibid and Mann and Plummer, *The Aspirin Wars*.

p.112 'Hugo Schweitzer was their cut-out . . .' Ambruster, *Treason's Peace*.

'Bayer wasn't the only . . .' Haynes, *American Chemical Industry*. See also F. Dyer, *Edison, his Life and Inventions* (New York: Harper & Bros, 1929).

'Two days later . . .' ibid; Ambruster, *Treason's Peace*; G. Simon Deposition (Entry 199), *Records of the Office of Alien Property*, US National Archives, Washington, DC. See also 'Aspirin and Espionage', *Journal of American Medical Association* (1919).

p.113 'In actual fact . . .' ibid.

'four and a half million pounds of explosives . . .' *New York Times*, 25 April 1919.

'The jubilation was short-lived . . .' H. Landau, *The Enemy Within* (New York: G. P. Putnam, 1937).

'leaked them to the press instead . . .' *New York World*, 15–19 August 1915.

p.114 'A short while later an embarrassed Thomas Edison . . .' Haynes, *American Chemical Industry*.

'When the New York police . . .' Mann and Plummer, *The Aspirin Wars*.

'The reader will not be urged . . .' 'Big campaign for aspirin to forestall expiration of patent', *Printer's Ink*, June 1916.

'Under the single headline . . .' Advertisement in *New York Times*, 2 October 1916.

p.115 'But this false modesty . . .' *Journal of the American Medical Association*, 20 January 1917.

p.116 'Aspirin' Trade Mark: ibid., 27 January 1917.

'In the three years since 1914 . . .' Haynes, *American Chemical Industry*.

'Germany, encouraged by a belief . . .' N. Grant, *Illustrated History of 20th Century Conflict* (London: Hamlyn, 1992).

'The office of the APC . . .' See A. Mitchell Palmer, *Aims and Purposes of the Chemical Foundation Inc . . .* (New York: De Vinne Press, 1919).

pp.117 'This agency, the Bureau of Investigation . . .' Diarmuid Jeffreys, *The Bureau – Inside the Modern FBI* (London: Macmillan, 1994).

p.117–18 'The years before the war . . .' Verg et al., *Milestones*.

Fate of heroin . . . (and footnote): *Sunday Times*, 13 September 1998.

p.118 'Carl Duisberg had watched . . .' H. Flechtner, *Carl Duisberg: vom Chemiker zum Wirtschaftsführer* (Econ. Düsseldorf, 1959); and Mann and Plummer, *The Aspirin Wars.*

p.119 'None the less, in 1904 he sat down . . . J. Beer, *The Emergence of the German Dye Industry* (Illinois Studies in Social Sciences, University of Illinois Press, 1959); P. Hayes, *Industry and Ideology. I.G. Farben in the Nazi Era* (Cambridge University Press, 1987).

p.120 'Duisberg realized that if these relationships could . . .' ibid.

Footnote: letter to Bauer quoted in J. Borkin, *The Crime and Punishment of I.G. Farben* (New York: Free Press, 1978).

p.121 Footnote (Aspirin and the Russian Revolution): see D. G. Friend, *Archives of Surgery*, vol. 108, 1974.

p.122 Footnote: 'On 9 September 1916 . . .' Borkin, *The Crime and Punishment of I.G. Farben.*

7: CIVILIZATION COULD DISAPPEAR . . .

p.123 At first sight . . .' These observations are based on a visit I made to Seaford Cemetery in East Sussex.

p.123 'I had a little bird . . .' This rhyme was popular across North America in the years of the epidemic. See R. Crawford, *The Spanish Flu, Stranger than Fiction: Vignettes of San Diego History* (San Diego Historical Society, 1995).

p.124 'It killed at least five times as many people as had perished in the war . . .' This may be a conservative figure. Some estimates range as high as 70–80 million. See notes for page 139.

'America's Selective Service Act . . .' Passed by Congress shortly after America's entry into the war in 1917. Levels of voluntary enlistment were actually quite high in the United States (given that it was for a war being fought overseas), but like other combatant nations it had no real alternative but to introduce conscription.

Fort Riley, Kansas: This was a huge establishment housing between 25,000 and 28,000 men, many in basic training or waiting for onward posting to their units. Crammed together in Spartan conditions, they endured sub-zero temperatures in winter and sweltering hot summers. See *Order of Battle of the United States Land Forces in The World War. Zone of the Interior: Territorial Departments. Tactical Divisions Organised in 1918. Posts, Camps and Stations.* (Washington, DC, Centre of Military History, US Army).

'Shortly before reveille . . .' See *The American Experience – Influenza 1918* (PBS, 1996) and Alfred W. Crosby, *Epidemic and Peace 1918: America's Forgotten Pandemic* (Cambridge University Press, 1976).

p.125 'When he finally got there . . .' ibid. and P. Brennan, 'The Great Dying: The 1918 Influenza Epidemic,' *Newsmax*, 31 October 2001.

'Thus by a few minutes . . .' Private Gitchell may have been the first *recorded* victim of the 1918 pandemic, but as there was no legal requirement in the United States at that time to report flu cases there may have been other earlier victims among the civilian population.

'Even the forty-eight fatalities . . .' *The American Experience – Influenza 1918* (PBS, 1996).

pp.125–6 'Later, of course, scientists would . . .' Gina Kolata, *Flu: The Story of the Great Influenza Pandemic of 1918 and the Search for the Virus that Caused It* (New York: Farrar, Straus and Giroux, 1999); P. W. Ewald, *Evolution of Infectious Diseases* (Oxford University Press, 1994).

p.126 'The 15th US Cavalry . . .' Richard Collier, *The Plague of the Spanish Lady* (New York: Atheneum, 1974).

p.127 'Germany's General Erich von Ludendorff . . .' ibid.

'The British Army's 29th Division . . .' S. Gillon, *The Story of the 29th Division* (London: Thomas Nelson & Sons, 1925).

'The 2nd Battalion . . .' *War Diary, 2nd Batt. Kings Royal Rifle Company 1914–1918*. See also 'Influenza Epidemic in the British Army in France, 1918. Influenza Committee of the Advisory Board to the D.G.M.S., France', *British Medical Journal*, 1918, and S. Cummins, *Studies of Influenza in Hospitals of the British Armies in France 1918* (Medical Research Council, Special Report Series no. 36, London, 1919)

'the whole epidemic being dubbed The Spanish Influenza'. See *British Medical Journal*, 13 July 1918 and 19 November 1918. Also Richard E. Shope, 'Old, intermediate and contemporary contributions to our knowledge of pandemic influenza', *Medicine*, vol. 23, 1944.

p.128 'The US Navy sailors . . .' Crosby, *Epidemic and Peace*.

'To Hungarians . . .' Collier, *The Plague of the Spanish Lady* and R. Neustadt, *Bacteriologie*, November 1928.

Footnote: see G. Rice, *Black November: The 1918 Influenza Epidemic in New Zealand* (London: Allen & Unwin, 1988).

p.129 'a life-and-death battle . . .' Crosby, *Epidemic and Peace*. 'Two hours after admission . . .' *British Medical Journal*, 22 December 1979.

'Alarmed by reports . . .' Kolata, *Flu*. See also Simon Flexner and James T. Flexner, *William Henry Welch and the Heroic Age of American Medicine* (Baltimore: Johns Hopkins University Press, 1941) and Victor C. Vaughan, *A Doctor's Memories* (Indianapolis: Bobbs Merrill, 1926).

p.130 'it was spreading like a bush fire . . .' Crosby, *Epidemic and Peace*.

Newfoundland: *Evening Telegram*, 30 September 1918; *Daily News*, 1 October 1918; *Evening Telegram*, 9 October, 1918.

Glasgow: *British Medical Journal*, October 1918.

Lagos: Collier, *The Plague of the Spanish Lady*.

Archangel: ibid.

Rhodesia: ibid.

Letterkenny: ibid.

pp.130–1 Tasmania: ibid.

p.131 'The story of how it hit Fiji and New Zealand . . .' (and footnote p.132): Rice, *Black November*.

p.132 'Yesterday there were . . .' *The Times*, 24 October 1918.

p.133 'Shortly after the virus's . . .' Crosby, *Epidemic and Peace*. 'Another rumour . . .' Collier, *The Plague of the Spanish Lady*.

'Phillip Doana . . .' *Philadelphia Inquirer*, 21 September 1918.

'yet ten days later, 200,000 people . . .' *The American Experience – Influenza 1918* (PBS, 1996); and death toll: G. F. Pyle, *The Diffusion of Influenza: Patterns and Paradigms* (New Jersey: p.334: Roman & Littlefield, 1986).

p.134 'In New York . . .' *New York Times*, 30 September 1918.

'A couple of weeks later . . .' New York State Department of Health, *A Special Report on the Mortality from Influenza during the Epidemic of 1918–19* (1923).

'The American Public Health Association issued . . .' *Journal of American Medical Association*, 4, 10 and 21 December 1918.

'Huge drafts of soldiers . . .' Crosby, *Epidemic and Peace*.

'every town dweller . . .' *British Medical Journal*, 21 December 1918.

p.135 'wash inside your nose . . .' *News of the World*, 3 November 1918.

'A Dr C. Y. White of Philadelphia . . .' Collier, *The Plague of the Spanish Lady*.

pp.135–6 'If the epidemic . . .' *The American Experience – Influenza 1918* (PBS, 1996).

p.136 'In October, the US Surgeon General, Rupert Blue . . .' *New York Times*, October 1918.

'It wasn't totally unique . . .' 'Aspirin für "La Grippe"', *Apotheker Zeitung*, 1920.

pp.136–7 'When no official "cure" . . .' alternative remedies (and footnote): see Collier, *The Plague of the Spanish Lady*.

p.137 'although a Philadelphian doctor . . .' *Journal of American Medical Association*, November 1918.

'In Delhi . . .' Collier, *The Plague of the Spanish Lady.*

'In London . . .' *Lancet*, November 1918.

'A fellow countryman . . .' *Daily Mail*, 10 December 1918.

p.138 'When influenza reached Australia . . .' R. Grenville-Smith and A. Barrie, *Aspro – how a family business grew up* (Nicholas International Ltd, 1976).

'In New Zealand . . .' Collier, *The Plague of the Spanish Lady.*

'In Camp Sherman . . . 100,000 aspirin tablets' ibid.

'In Paris . . .' *Le Monde*, 18 October 1918.

pp.138–9 Roosevelt to Admiral Dot: Collier, *The Plague of the Spanish Lady* – except Lloyd George: A. Grimes, *Manchester Evening News,* November 1998; the Kaiser: A. A. Hoehling, *The Great Epidemic* (Boston, Mass.: Little Brown, 1961); and Woodrow Wilson: D.J. Tice, *Flu Deaths* (Pioneer Planet, 1997).

p.139 'And for every supposedly . . .' For a cross-section of international and localized fatality estimates see Crosby, *Epidemic and Peace;* S. Collins and J. Lehman, *Excess Deaths from Influenza and Pneumonia* (Public Health Monographs 1953); Gerard F. Pyle, *The Diffusion of Influenza: Patterns and Paradigms;* Kolata, *Flu*; Fred R. Van Hartesveldt, *The 1918–1919 Pandemic of Influenza* (Edwin Mellen Press, 1993); 'The great Ethiopian influenza epidemic of 1918', *Ethiopian Medical Journal*, vol. 27, (1990); *The American Experience – Influenza 1918* (PBS, 1996); Rice, *Black November*; Molly Billings, *The Influenza Pandemic of 1918*, see website at www.stanford/edu/group/virus; W. I. Beveridge, *Influenza: The Last Great Plague* (Prodist, 1977); Collier, *The Plague of the Spanish Lady;* E. Brainerd and M. Siegler, *The Economic Effects of the 1918 Influenza Epidemic* (Centre for Economic Policy Research, 2003).

'How that happened . . .' Crosby, *Epidemic and Peace.*

p.140 'The young Canadian . . .' Seaford Cemetery contains 253 war graves of which 191 are Canadian (a base on the outskirts of the town was one of the Canadian Army's main training centres in the First World War and thousands of troops passed back through it prior to demobilization). While many of them died as a result of wounds received in battle, many others clearly died of the flu. For information on one of those victims, Sapper Charles Emerson McCallum, see the website dedicated to him at www.members.aol.com/_ht_a/reubique/3136830. For Canadian deaths generally see *Saskatchewan History*, vol. 49, 1997 and J. P. D. McGinnis, *The Impact of Epidemic Influenza in Canada* (Medicine in Canada, Historical Perspectives, 1981).

pp.141–2 'Production and sales of aspirin more than doubled . . .' Figures from 1899 to 1944 for Bayer's international and domestic markets are available in table form at the company's Leverkusen archives. They show a rise from 71,186 kilos in 1918 to 150,260 by 1919 before dropping back to 70,046 kilos by 1921 when the epidemic was over. These figures are particularly striking when one considers that production at Leverkusen was otherwise severely hampered by events at the end of the war. Consumption of aspirin in the UK rose by 92 per cent (figures in pounds weight unavailable) between 1918 and 1920 (Pharmaceutical Society, 1922). Demand for Aspro in Australia and New Zealand increased to such an extent that the Nicholas brothers were able to expand overseas as a result. Sales exploded again when a second (but milder) flu epidemic broke out in the late 1920s: see Grenville-Smith and Barrie, *Aspro.*

p.142 'the Aspirin Age . . .' This phrase was later popularized following the publication of a book by the same name: Isabel Leighton, *The Aspirin Age* (London: Bodley Head, 1950). In the book's preface she wrote, 'During these throbbing years we searched in vain for a cure-all (for the headaches of the world situation), coming no closer to it than the aspirin bottle.'

8: THE ASPIRIN AGE

p.143 'For Carl Duisberg . . .' E. Verg, G. Plumpe and H. Schultheis, *Milestones* (Bayer AG, 1988); L. F. Haber, *The Chemical Industry, 1900–1930: international growth and technological change* (Oxford: Clarendon Press, 1971).

p.144 'The situation in the United States was particularly galling . . .' *Records of the Office of Alien Property, 131* (US National Archives, Washington, DC).

'Sterling Products . . .' (and its purchase of Bayer): J. Hiebert, *Our Policy Is People: Their Health Our Business* (New York: The Newcomers Society, 1963); *Drug and Chemical Markets,* 18 December 1918; C. Mann and M. Plummer, *The Aspirin Wars, Money, Medicine and 100 years of Rampant Competition* (New York: Knopf, 1991). T. N. Reimer, *Bayer & Company in the United States: German dyes. Drugs and cartels in the progressive era* (PhD. thesis, Syracuse University, 1996).

p.146 'the field has been merely scratched . . .' Sterling, Number 10, Department of Justice Central Files, Case 60–21–56 (Sterling Products, Inc) RG 60 (US National Archives, Washington, DC).

'When all was said and done . . .' Mann and Plummer, *The Aspirin Wars.*

'erroneous and misleading . . .' *Farbenfabriken vormals Friedrich Bayer & Co v. Chemische Fabrik Von Heyden* (Reports of Patent, Design and Trade Mark Cases, 1905, 22: 501–18).

'He had kept his head down . . .' W. Korthaus, *Pharmazeutische Geschäft in Südamerika währen des Krieges* (Bayer Leverkusen archives).

p.147 'At the end of April 1919 . . .' A. M. Luckau, *The German Delegation at the Paris Peace Conference* (New York: Columbia University Press, 1941); A. Sharp, *The Versailles Settlement – peacemaking in Paris 1919* (London: Macmillan, 1991).

'Among the delegates . . .' J. Borkin, *The Crime and Punishment of I.G. Farben* (New York: Free Press, 1978).

'The final document . . .' A. J. P. Taylor, *Origins of the Second World War* (London: Hamish Hamilton, 1961).

p.148 'For Farbenfabriken Bayer . . .' Haber, *The Chemical Industry, 1900–1930.*

Duisberg meeting with Weiss: *Wiederschrift über die Besprechung am Montag, den 22 September 1919 mit William Weiss et al* (Bayer Leverkusen archives).

'Not surprisingly, the talks were inconclusive . . .' Mann and Plummer, *The Aspirin Wars.*

p.149 'Everywhere in the whole world . . .' *Bericht über die Konferenz mit Herrn Weiss aus New York vom 8 April 1920* (Bayer Leverkusen archives).

'An interim agreement was reached on 28 October 1920 . . .' *Exhibit B: Agreement between The Bayer Company*. . . . US v Alba Pharmaceutical Company et al. (Trade Cases, 1941).

'But finally, on 9 April 1923, . . .' *Contract between Farbenfabriken Bayer and Winthrop Chemical Company Apr. 9, 1923 Exhibit A,* US v Alba; *Contract between Farbenfabriken Bayer and the Bayer Company 9 April 1923, Exhibit A,* US v The Bayer Company et al. (Trade Cases, 1941).

p.150 Weimar economy: C. Bresciani-Turroni, *The Economics of Inflation* (London: Allen & Unwin, 1937).

p.151 'One of the Bayer . . .' *Bayer Company v United Drug Company* (Federal Reporter, 1921); *United Drug Company v Farbenfabriken of Elberfeld* (US Patent Office Cancellation No. 424)

'Therefore for sales of ASA to them . . .' *Drug. Chem. Markets* 1921 and J. McTavish, 'What's in a name? Aspirin and the American Medical Association', *Bulletin of Historical Medicine*, 1987.

p.152 'In Canada, for example, . . .' ibid.

p.153 'George Davies was enjoying life . . .' R. Grenville-Smith and A. Barrie, *Aspro – how a family business grew up* (Nicholas International Ltd, 1976)

p.154 'Chemists Give Away . . .' *The Queenslander*, August 1918.

'An Aspro tablet . . .' *Melbourne Herald*, November 1920.

p.155 'At other times his copy . . .' Grenville-Smith and Barrie, *Aspro*; and B. Morgan, *Apothecary's Venture: The Scientific Quest of the International Nicholas Organisation* (Nicholas Kiwi, 1959).

'This is the first time . . .' *The Australian*, July 1919.

p.156 'They opened up a factory . . .' Morgan *Apothecary's Venture.*

British aspirin brands: see *British Pharmacopoeia* for years 1906 to 1930.

'Its American antagonist and soon-to-be-partner . . .' *Sterling-Winthrop Group Ltd v Farbenfabriken Bayer A.G.* (Reports of Patent Cases 1976).

p.157 'Their first attempt was a disaster . . .' Grenville-Smith and Barrie, *Aspro.*

'passed to a London agency . . .' Samson & Clark, which worked for Golling & Co, the British distributors of Aspro (*Chem & Drug*, 1927).

p.158 'The advertisements were masterpieces of salesmanship . . .' See *Hull Daily*

Mail throughout 1926–7 (7, 14, 21, 28 January, 25 February, 8 March, 29 April, etc.).

p.161 'One market remained unchallenged . . .' Grenville-Smith and Barrie, *Aspro.*

'Most consisted . . .' *Delineator*, December 1921.

'This led to ever more imaginative . . .' *Federal Trade Commission Decisions* (*924*–1935; *921*–1936, *210*–1936).

p.162 'One of the first of these . . .' Mann and Plummer, *The Aspirin Wars.*

'Eventually the advertising . . .' I. Brichta, *The Promise and the Product: 200 years of American advertising posters* (London: Macmillan, 1979).

'But it was through radio . . .' H. Hettinger, *A Decade of Radio Advertising* (University of Chicago Press, 1993); S. Fox, *The Mirror Makers: a history of American advertising and its creators* (New York: Morrow, 1984).

p.163 'The often exaggerated claims . . .' See the following Federal Trade Commission rulings: a) 19–229–1936; b) 20–695–1935; c) 24–115–1936; d) 21–924–1935; e) 22–921–1936; f) 23–1124–1935, etc.

'As the English humorist . . .' J. K. Jerome, *Three Men in a Boat*, 1889 (London: Bloomsbury Classics, 1997).

p.164 'Jose Ortega y Gasset . . .' *Revolt of the Masses* (Paris: Wood, 1931).

Kafka and Caruso: see Verg et al., *Milestones.*

G. Orwell: *The Road to Wigan Pier* (London: Victor Gollancz, 1937).

G. Greene: *Stamboul Train* (London: Heinemann, 1932).

Edgar Wallace: *Door with Seven Locks* (Leipzig: Tauchnitz, 1926).

'Counterfeiting . . .' Verg et al., *Milestones.*

p.165 'Some criminals took the more direct route . . .' *New York Times*, 17 December 1927.

'The District Attorney . . .' *New York Times*, 17 January 1928.

'This ran parallel to the mistaken theory . . .' A. Eichengrün: '50 Jahre Aspirin', *Pharmazie*, 1949.

'Dr R. Eccles Smith . . .' *British Medical Journal*, August 1920.

p.166 'Dr Ethelbert Hearn . . .' ibid., 11 September 1920.

'H.E. Davidson . . .' ibid., 22 September 1928.

'The details of one . . .' *Daily Sketch*, 10 July 1929, and *British Medical Journal*, 20 July 1929.

'Sir Robert Thomas MP . . .' *Hansard*, 13 February 1929.

'Dr Gerald Stot . . .' *Lancet*, 6 June 1931.

'Not the least of the contributions . . .' *Lancet* 14 September 1935.

p.167 'But things were changing at Leverkusen . . .' *Fifty Years of Bayer Remedies* (Leverkusen, 1938).

9: A MORAL COLLAPSE

p.168 'Farbenfabriken Bayer emerged . . .' C. Mann and M. Plummer, *The Aspirin Wars, Money, Medicine and 100 Years of Rampant Competition* (New York: Knopf, 1991)

p.169 Formation and early days of IG Farben: J. Borkin, *The Crime and Punishment of I.G. Farben* (New York: Free Press, 1978); L. F. Haber, *The Chemical Industry 1900–1930: international growth and technological change* (Oxford: Clarendon Press, 1971); P. Hayes, *Industry and Ideology: I.G. Farben in the Nazi Era* (Cambridge University Press, 1987); E. Verg, G. Plumpe and H. Schultheis, *Milestones* (Bayer AG, 1988).

p.170 'The IG's association . . .' *Trials of the War Criminals Before the Nuremberg Military Tribunals under Control Council Law 10* (Imperial War Museum and PRO: FO 646); Hayes, *Industry and Ideology*.

p.171 'A few days later Bosch . . .' Hayes, *Industry and Ideology*.

'the cash-rich National Socialists . . .' After Hitler won the election, IG Farben gave the Nazis another 100,000 marks. ibid.

'Bosch was never the most enthusiastic . . .' (and footnote): Verg et al., *Milestones*.

'The combine had invested . . .' Mann and Plummer, *The Aspirin Wars*.

p.172 'As history now knows . . .' *Trials of the War Criminals Before the Nuremberg Military Tribunals;* Borkin, *The Crime and Punishment of I.G. Farben*. See also *The Judgement in the Farben Trial* (Bollwerk-Verlag Karl Drott 1948).

p.173 'It produced, through its Degesch . . .' The question of whether or not IG Farben had a direct role in the production of Zyklon B has been the subject of much controversy since the war. The judges at Nuremberg found that while the IG had a stake in Degesch (Deutsche Gessellschaft für Schadlingsbekamp-fung) it did not have any persuasive influence on the management of the company. They were therefore unable to resolve the matter conclusively and ruled the charge unproven. Many historians have found this hard to accept and I agree with them. At least one senior IG executive had a seat on the board and given the IG's close association with Nazis at Auschwitz and at the nearby IG plant, IG Monowitz, it stretches credibility beyond breaking point to suggest that its managers remained completely ignorant of what was going on a few hundred yards away. This, however, is the version of events contained in all Bayer's official histories.

'An obituary . . .' *The Times*, March 1935, written by Henry Armstrong, a leading British scientist, and reprinted around the world.

'He had to make some concessions . . .' *Contract between I.G. and Winthrop Chemical Co., Exhibit A* US v Alba, 15 November 1926; *Contract between I.G. and The Bayer Company Inc., Exhibit A,* US v Bayer, 15 November 1926.

p.174 'He gave Max Wojahn . . .' Mann and Plummer, *The Aspirin Wars*.

'It was called Cafiaspirina . . .' Numerous examples of advertising for this product can be found in Bayer Leverkusen archives.

'They travelled in specially . . .' See *Farbenfabriken Bayer AG v Sterling Drug Inc.* District Court New Jersey (18 February 1960) in affidavit given by W. Kronstein; and *Merchants in Mexico* (Central America Trade Letters No. 17, 1936). Footnote: Grenville-Smith and Barrie, *Aspro.*

p.175 'The returns were exceptional . . .' *Aufstellungen über an I.G. und GAW ausgeschutte Gewinne 1919–37* (Bayer Leverkusen archives).

'Throughout the 1930s much . . .' Hayes, *Industry and Ideology.*

'However, Mann was also a Nazi supporter and had consistently . . .' Wilhelm Mann joined the Nazi Party in 1933. Sterling, Number 10, Department of Justice Central Files, Case 60–21–56 (*Sterling Products, Inc.*) RG 60 (US National Archives, Washington, DC).

'Alarm bells should have rung . . .' Mann and Plummer, *The Aspirin Wars.*

'Matters got even worse . . .' ibid.

p.176 'The only way to get . . .' ibid.

'In May 1941 . . .' *New York Herald Tribune*, 29 May 1941; *PM*, 1 June 1941.

p.178 'With a heavy heart . . . Cable from William Weiss to Wilhelm Mann, 15 August 1941 and reply from Vorstand Farbenindustrie to Weiss, 26 August 1941 in Bayer Leverkusen archives.

'died in a road accident . . .' Mann and Plummer, *The Aspirin Wars.*

'The Hall of Honour . . .' Many of the personal and biographical details about Arthur Eichengrün in these pages are drawn from information generously provided (via documents and telephone interviews) by his grandson Ernst Eichengrün and are the result of his own careful and detailed research into his family history. Where it has been possible and necessary to verify this information or to cross-reference it with other sources (such as Arthur Eichengrün's own writings) I have done so and have cited them below. However, the narrative interpretation I have placed on these events is largely my own.

p.179 'Inside the first case . . .' A. Eichengrün, '50 Jahre Aspirin', *Pharmazie*, 1949.

'When Eichengrün joined . . .' Hans Schadewaldt and Rosemary Alstaedter, *History of Pharmacological Research at Bayer* (Bayer, 1991).

p.180 'What perhaps is more relevant . . .' Eichengrün, '50 Jahre'.

'to the round robin discussion . . .' Aspirin's 'baptism certificate' (Bayer Leverkusen archives).

As for credit . . .' *Daily Mail*, 11 May 1920, *Chemical Trades Journal* (2), 1929, and *Handbuch der deutschen Gesellschaft*, 1930.

p.181 'After all, the unsuccessful German patent . . .' Ernst Eichengrün. Incidentally, Bonhoeffer's name also appeared on a British patent for various other salicylic acid derivatives. See *British Patent 9123*, 3 March 1900. Arthur Eichengrün's post-Bayer business career: Ernst Eichengrün. See also

Stadlinger H. Gedetktage, 'Arthur Eichengrün 80 Jahre', Pharmazie, 1947; E. Escales, *A. Eichengrün 80 Jahre* (Kunstoffe, 1947); H. G. Bodenbender, 'A. Eichengrün zum 80 Geburstag', *Angewandte Chemie,* 1948.
'As a result he began . . .' Ernst Eichengrün.

p.183 Effect of Nazi Jewish laws on Arthur Eichengrün and his business: Ernst Eichengrün.
Goering anecdote: ibid.
'In 1934, a recently retired . . .' A. Schmidt, *Die industrielle Chemie in ihrer Bedeutung im Weltbild und Erinnerungen an ihren Aufban* (Berlin: De Greuter, 1934).

p.184 'Heinrich Dreser, on the other hand . . .' Ernst Eichengrün.
Footnote: A. Eichengrün, *Pharmaceutisch-wissenschafliche Abteiling.* In: *Geschichte und Entwicklung der Farbenfabriken vorm Friedr Bayer & Co, Elberfeld, in den ersten 50 Jahren* (Munich: Meisenbach-Riffrath, 1918). See notes for page 290.

p.185 'Whatever the case, the moment passed . . .' Ernst Eichengrün.
'Then one of the Nazi . . .' Unknown but probably *Volkischer Beobachter*. For more on Nazi Jewish persecution of 1938 see A. Read, *Kristallnacht – unleashing the holocaust* (London: Michael Joseph, 1989).
'Somehow he got by . . .' Ernst Eichengrün.

p.186 'Ironically, it was his wife . . .' ibid. Eichengrün was formally denounced by the President of the German Patent Office.
'The regime at Theresienstadt . . .' see www.remember.org/chuck and G. E. Berkley, *Hitler's Gift: The story of Theresienstadt* (Boston, Mass.: Branden Books, 1993).
Eichengrün at Theresienstadt: Ernst Eichengrün.

p.187 'He wrote a letter . . .' *Dr A. Eichengrün, Aspirin, KZ Theresienstadt 1944* (Bayer Leverkusen archives). This letter is particularly notable for a sentence that follows his description of the lack of credit given to him by the Deutsche Museum Hall of Honour for his role in the discovery of aspirin. It reads, 'To what influences this omission is to be attributed, can only be assumed.' Under the circumstances, this was a brave thing to write.
'On Heinrich Himmler's orders . . .' Berkley, *Hitler's Gift*. 'When Eichengrün finally managed to get back to Berlin . . .' Ernst Eichengrün.

p.188 'It was published . . .' Eichengrün, '50 Jahre'.
'and would continue to maintain . . .' e.g. letter from a Dr Rempen of Bayer AG to a Dr C. Fouquey, College de France, Paris, 7 July 1982, which states clearly that Eichengrün was 'not involved in the origin of Aspirin', (Bayer Leverkusen archives).
'It wasn't until Dr Walter Sneader . . .' W. Sneader, 'The discovery of aspirin: a reappraisal', *British Medical Journal*, 23 December 2000. The article generated a heated exchange of letters between Bayer AG and Dr Sneader in

subsequent editions of the same journal: letters to *BMJ*, 30 January 2001; 1 February 2001; 9 and 16 March 2001.

p.189 'By the mid 1940s . . .' Borkin, *The Crime and Punishment of I.G. Farben;* Hayes, *Industry and Ideology*.

'To take just three . . .' Hayes, *Industry and Ideology*.

'One tragic story . . .' This account of the experiences of the Mozes twins at Auschwitz was told to me by Eva Mozes Kor. Her full story can also be read in her autobiography, *Echoes from Auschwitz* (Candles, 1999).

p.190 'It was because . . .' L. M. Lagnado and S. Cohn Dekel, *Children of the Flames: Dr Josef Mengele and the untold story of the twins of Auschwitz* (London: Sidgwick and Jackson, 1991).

'Over the following months . . .' Eva Mozes Kor.

'batch number BE 1034 . . .' ABC 20/20 (ABC News 2000).

'After one set of injections . . .' Eva Mozes Kor.

'As Wilhelm Mann . . .' ABC 20/20 (ABC News 2000). Bayer AG's response to this letter is that there is no evidence the money was ever sent.

p.191 'the nearby IG Monowitz . . .' See Bernd C. Wagner, *IG Auschwitz, Zwangsarbeit und Vernichtung von Haftlingen des Lagers Monowitz 1941–1945* (Munich: K.G. Saur, 2000).

'including Dr Helmuth Vetter . . . 1943 he conducted research . . .' Baruch C. Cohen, *The Ethics of Using Medical Data from Nazi Experiments* (Jewish Law: Articles 1997–2003).

'I have thrown myself . . .' ABC 20/20 (ABC News 2000).

'Emotionally I have forgiven . . .' Eva Mozes Kor. Her parents, grandparents, two older sisters, uncles, aunts and cousins were all killed in the Holocaust.

p.192 'As for the many . . .' *Trials of the War Criminals Before the Nuremberg Military Tribunals under Control Council Law 10*.

'Concentration camp prisoners . . .' ibid.

'After Nuremberg, the IG Farben . . .' Verg et al., *Milestones*.

'In 1956, not long after . . .' ibid.

BOOK THREE

10: SOLUBLE SOLUTIONS AND COSTLY COMPETITION

p.195 'It was a dismal day to arrive . . .' The story of the invention and development of soluble Disprin contained in these pages is based on documents held at the Reckitt's Heritage archives of Reckitt Benckiser in Hull. Chief among these is a fascinating first-hand account written by George Colman Green in 1969. As far as I have been able to determine, his paper, *The Archaeology and Social History of Disprin*, has never been published outside the company – although I understand a version may have seen the light of day in the company's *Reckitt and Colman Magazine* around that time. Other sources are cited where necessary.

'At the beginning of 1938 . . .' *British Medical Journal*, 1938.

p.196 'The following year, another eminent physician . . .' *Lancet*, 1939.

p.197 'Just before the war . . .' George Colman Green, *The Archaeology and Social History of Disprin.*

'Then one night . . .' *Hull Daily Mail*, 19 July 1941.

'One of those whose work was lost . . .' George Colman Green.

'Scruton was well aware . . .' ibid.

p.198 'although as the formula . . .' By the start of the Second World War the chemical composition of acetylsalicylic acid and the process for its manufacture were being taught to first-year chemistry undergraduates.

p.199 'On 24 February . . .' Cited by George Colman Green as HAS/APB (24 February 1944).

'At a meeting . . .' ibid: Minutes of Hull Research Committee, 14 March 1944.

p.200 'And then someone remembered . . .' Profile of George Colman Green, 'Ours' in *Reckitt and Colman Magazine*, November 1969.

'It became obvious to me . . .' George Colman Green.

p.201 'The news . . . was accepted with incredulity . . .' ibid.

p.202 'based on a design . . .' Colman had been a member of one of the British Industrial Objectives Survey Export Teams. Set up in 1944, their aim was to find and safeguard enemy military and industrial technology before it was destroyed by the retreating German forces or snatched up by Britain's allies.

pp.202–3 'his long-suffering assistant . . .' Fred Dook was a long-standing Reckitt and Colman employee who had served with the RAMC in Malta and had taken part in the Normandy landings.

p.203 'In the meantime . . .' George Colman Green.

'There were many other obstacles to overcome . . .' B. Reckitt, *The History of Reckitt and Sons* (London: A. Brown & Sons Ltd, 1952).

'quiet game of snooker . . .' 'Ours' in *Reckitt and Colman Magazine*, November 1969.

p.204 'So many technical problems . . .' George Colman Green.

'The response was everything . . .' B. Reckitt, *The History of Reckitt and Sons.*

'I have often wondered . . .' George Colman Green.

p.205 'Its "Take an Aspirin – I mean a Disprin" slogan . . .' 'The history and development of the Reckitt analgesic preparations', *Reckitt and Colman Magazine*, 1962.

pp.205–6 Bayer Ltd and Sterling: Mann and Plummer, *The Aspirin Wars.*

p.206 'Before, during and after . . .' ibid.

'Some of these . . .' Hans Schadewaldt and Rosemary Alstaedter, *History of*

Pharmacological Research at Bayer (Bayer, 1991).

'Anacin, for instance . . .' *Wall Street Journal*, 14 March 1963.

p.207 'Another, Excedrin . . .' ibid.

'This had been first synthesized . . .' H. N. Morse, *Darstellungsmethode der Aceytlamidophenole* in *Berichte der Duetchen chemischen Gesellschaft* (1878).

'It reappeared in 1946 . . .' D. Lester and L. A. Greenberg, 'The metabolic fate of acetanilide', *Journal of Pharmacology and Experimental Therapeutics*, 1947.

'Two years later . . .' B.B. Brodie and J. Alexrod, 'The fate of acetanilide, ibid., 1949.

'Bayer Ltd's managing director . . .' Mann and Plummer, *The Aspirin Wars*.

'Reckitt and Colman had based . . .' 'The History and Development of the Reckitt Analgesic Preparations', *Reckitt and Colman Magazine*, 1962.

p.208 'This wasn't much of a problem in Britain . . .' Mann and Plummer, *The Aspirin Wars*.

'it took the unusual step . . .' *London Gazette*, January 1963.

p.209 'By the early 1970s . . .' *Pharmacology*, October 1974.

History of American Home Products: *Fortune* magazine, April 1958. And of Anacin: See *William S. Merrell v Anacin Company* US Court of Customs and Patent Appeals (1938).

p.210 Anacin commercial: *Business Week*, 10 October 1959.

'Bufferin's promotional slogan . . .' *Life* magazine, January 1949.

'Four ingredients instead . . .' *Wall Street Journal*, 20 September 1961.

'It joined in the orgy of television . . .' *Saturday Evening Post*, 5 January 1957.

p.211 'In 1962, the Federal Trade Commission . . .' T. DeKornfeld, L. Lasanga and T. M. Frazier, *A Comparative Study . . .*, *Journal of American Medical Association*, 29 December 1962.

'The company immediately issued advertisements . . .' *Life* magazine, 25 January 1963.

'It followed this up by filing . . .' *Wall Street Journal*, 10 April 1972.

p.212 'This whimsical container . . .' L. G. Foster, *A Company That Cares* (New Brunswick: Johnson & Johnson, 1986). See also M. E. Bowden, 'Tylenol: over 50 years from laboratory shelf to medicine cabinet', *Chemical Heritage*, vol. 19, 2001.

'Indeed the response was so good . . .' Mann and Plummer, *The Aspirin Wars*.

'In the meanwhile . . .' Foster, *A Company That Cares*.

'In 1967, consumers were told . . .' *Advertising Age*, 25 November 1968.

p.213 'cleverer sales pitches . . .' Foster, *A Company That Cares;* Mann and Plummer, *The Aspirin Wars*.

'within a year . . .' *Advertising Age*, 15 August 1977.

'Its share slumped to barely 10 per cent . . .' See *Market Shares of Selected*

Analgesics: In the Matter of Sterling Drug: Federal Trade Commission (1983). *Decision.*

'Over the next three years . . .' *In the Matter of Bristol-Myers*, Federal Trade Commission (1983) *Decision; In the Matter of American Home Products,* Federal Trade Commission (1983) *Decision; In the Matter of Sterling Drug:* Federal Trade Commission (1983) *Decision.* (All at Federal Records Center, Maryland, US).

'We, Boots Pure Drug Company . . .' (UK Patent 971700).

History of Boots: C. Weir, *Jesse Boot of Nottingham* (Nottingham: Boots Company, 1994) and www.boots-plc.com/ history.

p.214 'He quickly became convinced . . .' For Dr Stewart Adams and the discovery and development of ibuprofen, see J. S. Nicholson, 'Ibuprofen' in *Chronicles of Drug Discovery* (London: J. Wiley, 1982); S. Adams, 'The discovery of Ibuprofen', *Chemistry in Britain*, December 1987.

p.216 'In 1974, the Upjohn Company . . .' *Business Week*, 1 February 1982.

'But the big breakthrough . . .' *Advertising Age*, 2 April 1984.

'When the Food and Drug Administration . . .' *Over-the-Counter Drugs, Establishment of a Monograph for OTC Internal Analgesic, Antipyretic and Antirheumatic Products* Federal Register (8 July 1977).

'When Tylenol was withdrawn from sale . . .' *Business Week*, 18 October 1982.

p.217 'Their misery was compounded . . . Reye's syndrome.' Mann and Plummer, *The Aspirin Wars.*

11: SO THAT'S HOW IT WORKS!

p.219 'In late November . . .' *Oil, Paint and Drug Reporter*, 12 December 1956; *Pharmacy International*, February 1957; *Chemical and Engineering News*, 18 November 1957.

'There is still much . . .' *Pharmacy International*, February 1957.

p.220 'no one since Bayer's Heinrich Dreser . . .' *Pharmakologisches über Apirin-Acetylsalicylsaüre* (Archiv für die Gesammte Physiologie, 1899).

'To the boy . . .' This account of Sir John Vane's early life, education and his subsequent interest in aspirin is drawn from an interview he gave me in July 2002 and other information provided by his office.

p.222 'In science the credit goes to the man . . .' Sir Francis Darwin, *Eugenics Review*, April 1914. Darwin (1848–1925) was a botanist and the son of Charles Darwin.

'If you prove to be . . .' H. O. J. Collier, 'The story of aspirin', *Discoveries in Pharmacology*, vol. 2: *Haemodynamics, Hormones and Inflammation*, edited by Paraham and Bruinvels (Elsevier, 1984).

'Harry O. J. Collier was born in 1912 . . .' Biographical details provided by his son, Professor Joseph Collier.

p.223 'One of the first . . .' ibid.

'In 1958, he began to experiment . . .' H. O. J. Collier and P. G. Shorley, *British Journal of Pharmacology*, vol. 15. 1960.

p.224 'Although intriguing . . .' ibid.

'Back in 1898 . . .' H. Dreser, *Pharmakologisches über Aspirin-Acetylsalicylsäure* (Archiv für die Gerammte Physiologie, 1899).

p.225 'I'm not really sure what they were doing . . .' Joseph Collier.

'The questions that needed to be answered . . .' H. O. J. Collier, *Scientific American*, vol. 209, 1963.

p.226 'Fortunately, Collier was able to recruit some . . .' Sir John Vane.

'It was then that . . .' Joseph Collier.

'That laboratory . . .' Sir John Vane.

'One of his notable achievements . . .' J. R. Vane, 'The use of isolated organs for detecting active substances in the circulating blood', *British Journal of Pharmacology*, 1964.

p.227 'He asked Vane to take her on . . .' Sir John Vane.

p.228 'They had been conducting . . .' ibid.

'When he read their paper . . .' P. J. Piper and J. R. Vane, *Nature*, 1969; and Joseph Collier.

p.229 'One weekend in April . . .' Sir John Vane.

'It was a genuine Eureka moment . . .' This is not a description that Vane would ever dream of applying to himself, but a number of other scientists I spoke to whilst researching this book see his 'intuitive leap' in these terms.

'As he hurried . . .' Sir John Vane.

p.230 'But what did it mean? . . .' J. R. Vane 'Inhibition of prostaglandin synthesis as a mechanism of action for aspirin-like drugs', *Nature*, 23 June 1971.

'We went out for dinner . . .' Sir John Vane.

'He was angry . . .' Joseph Collier.

'Vane was just as concerned . . .' ibid.

p.231 'It was accompanied . . .' J. B. Smith and A. L. Willis, 'Aspirin selectively inhibits prostaglandin production in human platelets', *Nature*, 23 June 1971; S. H. Ferreira, S. Moncada and J. R. Vane, 'Indomethacin and aspirin abolish prostaglandin release from spleen', *Nature*, 23 June 1971.

'It was established . . .' J. R. Vane and R. M. Botting, 'Anti-Inflammatory drugs and their mechanism of action' (Inflammation Research, 1998).

'However, for the purpose of this narrative . . .' Many hundreds, if not thousands, of articles have been written on research into the various aspects of aspirin's mechanism of action. To list them all here would be impossible but detailed scientific bibliographies on the subject can be accessed via the Internet at such sites as www.ncbi.nlm.nh.gov (Pub Med: National Library of Medicine) or www.library.wellcome.ac.uk (Wellcome Library for the History and Understanding of Medicine).

p.232 'although some scientists . . .' see e.g. S. Abramson and G. Weissman, *Arthritis and Rheumatism* (January 1989); G. Weissman (Journ. Lipid. Mediat., 1993).

'Platelets . . .' P. Elwood and C. Hughes, 'A history of platelets, aspirin and cardiovascular disease', *Aspirin and Cardiovascular Disease*, University of Wales, 1997.

'In 1975, as part of the . . .' M. Hamberg, 'Thromboxanes: A new group . . .', *Proceedings of the National Academy of Sciences*, August 1975.

p.232 Footnote: See Roy Porter, *The Greatest Benefit to Mankind: A Medical History of Humanity from Antiquity to the Present*. (London: HarperCollins, 1997).

p.233 'It would have a number of consequences . . .' *Guardian*, 12 October 1982.

'His one-time assistant . . .' Sir John Vane.

'Sadly Harry Collier . . .' Joseph Collier.

12: AFFAIRS OF THE HEART

p.235 'A hundred years or so ago . . .' See P. Fleming, *A History of Cardiology* (Amsterdam: Rodopi, 1997); J. D. Howell, *Concepts of Heart-Related Diseases* in *The Cambridge World History of Human Disease* (Cambridge University Press, 1993); W. F. Bynum et al., *The Emergence of Modern Cardiology* (London: Wellcome Institute, Medical History 1985).

p.236 'For many years . . .' (and footnote): Porter, *The Greatest Benefit to Mankind*.

'Scientists thought that if they could . . .' D. Murray et al., *Surgery*, August 1937.

p.237 'The two Oxford scientists . . .' J. D. F. Poole and J. E. French, 'Thrombosis', *Journal of Atherosclerosis*, August 1961.

'Craven was an MD . . .' *Los Angeles Times*, 19 August 1957; *American Mercury*, March 1959.

'Thus it was that in 1950 he wrote . . .' L. L. Craven, 'Acetyl salicylic acid: possible prevention of coronary thrombosis', *Annals of Western Medicine and Surgery*, 1950. See also L. L. Craven, 'Experiences with aspirin in the non-specific prophylaxis of coronary thrombosis', *Mississippi Valley Medical Journal*, 1953.

'This had sparked his . . .' ibid.

p.238 'As Craven revealed in a subsequent paper . . .' L. L. Craven, 'Prevention of coronary thrombosis and cerebral thrombosis', *Mississippi Valley Medical Journal*, 1956. Footnote: *American Mercury*, March 1959.

p.239 'John O'Brien was born . . .' Personal details from Peter Elwood who wrote an obituary on O'Brien's death in October 2002.

'his invention in the early 1960s . . .' J. R. O'Brien, 'Platelet aggregation: some results from a new method of study', *Journal of Clinical Pathology*, 1962. Footnote: G. Born, 'Aggregation of blood platelets by adenosine diphosphate and its reversal', *Nature*, 1962.

'And then, following up similar experiments . . .' H. J. Weiss et al., 'The effect of the salicylates in the hemostatic properties of platelets in man', *Journal of Clinical Investigation*, 1968.

p.240 'most famously established . . .' R. Doll and A. B. Hill, *British Medical Journal*, 30 September 1950.

'It took four years of repetitive house jobs . . .' This account of Peter Elwood's remarkable contribution to the development and use of aspirin as a preventative treatment for heart attack is based on interviews he gave me in September/October 2002, documents he was subsequently kind enough to pass on, and his own published work. Other sources are cited where necessary.

p.242 'Things had changed at Nicholas . . .' R. Grenville-Smith and A. Barrie, *Aspro – how a family business grew up* (Nicholas International Ltd, 1976)

p.246 'By the time the trial ended . . .' P. C. Elwood, 'A randomised, controlled trial of acetyl salicylic acid in the secondary prevention of mortality from myocardial infarction', *British Medical Journal*, March 1974; 'Regular Aspirin Intake', Boston Collaborative Drug Surveillance Group, ibid.

p.247 Cardiff 2 Trial: P. C. Elwood and P. M. Sweetnam, 'Aspirin and secondary mortality after myocardial infarction', *Lancet*, 22 December 1979.

'The result was the . . .' AMIS Research Group, 'A randomised controlled trial in persons recovered from myocardial infarction', *Journal of the American Medical Association*, 15 February 1980.

p.250 'And then, in May 1980, an editorial . . .' *Lancet*, 31 May 1980.

'The man behind this radical reassessment . . .' Profile in *Oxford Today*, 1989.

Methodology of meta-analysis: R. Peto, *British Journal of Cancer*, 1976, and interview with Sir Richard Peto.

p.251 'Like many others . . .' ibid.

'In 1978, with Sir Richard Doll . . .' R. Peto et al., 'Randomised trial of prophylactic daily aspirin in British male doctors', *British Medical Journal*, 1988.

Footnote: Mann and Plummer, *The Aspirin Wars*.

p.252 'But there was still . . .' Peter Elwood.

p.253 'Sterling decided . . .' *Food, Drug and Cosmetic Report*, March 1983.

'But then, in March 1983 . . .' Food and Drug Administration: *Minutes and Transcript, Cardiovascular and Renal Drugs Advisory Committee, 1 March 1983* (General Records, Rockville, Maryland).

'Although reluctant . . .' Peter Elwood.

'The first witness . . .' *Minutes and Transcript, Cardiovascular and Renal Drugs Advisory Committee.*

p.254 'Then Peter Elwood took over . . .'

p.254 'You have to understand . . .' Elwood.

'He clearly anticipated . . .' ibid.

p.255 'They really attacked me . . .' ibid.

'One board member . . .' *Minutes and Transcript, Cardiovascular and Renal Drugs Advisory Committee.*

'Things had degenerated . . .' ibid.

p.256 'In the process . . .' Sir Richard Peto.

'They heard first . . .' Food and Drug Administration: *Minutes and Transcript, Cardiovascular and Renal Drugs Advisory Committee, 11 December, 1984* (General Records, Rockville, Maryland).

'The clinical viewpoint . . .' ibid.

p.257 'Just under a year later . . .' *Washington Post*, 11 October 1985.

13: A TWENTY-FIRST CENTURY WONDER DRUG

p.258 'On 6 March 1999 . . .' *Associated Press*, 6 March 1999; *Daily Telegraph*, 7 March 1999; *Die Welt*, 7 March 1999, etc. With this stunt Bayer AG also found its way into the *Guinness Book of Records.*

'The building stood . . .' Duisberg's villa was demolished in the 1950s.

p.259 'For more than forty years . . .' C. Mann and M. Plummer, *The Aspirin Wars, Money, Medicine and 100 Years of Rampant Competition* (New York: Knopf, 1991).

'Bayer managed to persuade the courts . . .' *Sterling-Winthrop Group Ltd v Farbenfabriken Bayer A.G.* (Reports of Patent Cases, 1976).

'By buying Miles Laboratories Inc. . . .' *Chemical and Engineering News*, December 1977.

p.260 'as leading companies embarked on a merger . . .' e.g. *Financial Times*, 27 July 1989; *Business Week,* September 1993.

'struck a deal with Eastman Kodak . . .' *Financial Times*, 29 August 1994.

'This flag-waving was too much . . .' *Chemical and Engineering News*, September 1994 and December 2002.

'Because the ratio . . .' P. Elwood and C. Hughes, 'A history of platelets, aspirin and cardiovascular disease', *Aspirin and Cardiovascular Disease*, University of Wales, 1997.

p.261 'it was left to an American epidemiologist . . .' Physicians' Health Study Research Group: Preliminary Report, *New England Journal of Medicine*, January 1988.

p.262 'When the news broke . . .' Reuters, 26 January 1988; *BBC News*, 28 January 1988; *Newsweek*, 8 February 1988.

'but once they'd been put through . . .' Peto et al., 'Antiplatelet trialists collaboration', *British Medical Journal*, 1994; and Sir Richard Peto interview.

p.263 'One of the first people to look at aspirin . . .' Mann and Plummer, *The Aspirin Wars*. See also W. Fields and W. Hass, *Aspirin Platelets and Stroke, Background for a Clinical Trial* (Warren Green, 1971).

Footnote: One of the earliest of these prescribers of aspirin for stroke may have been Winston Churchill's private physician, who is believed to have given it to Britain's most famous Prime Minister from 1953. See 'Lord Moran's prescriptions for Churchill', *British Medical Journal*, June 1995.

p.264 'His study (the first part of which . . .' W. Fields et al., *Stroke*, May 1977.

'But a few months later another trial from . . .' Canadian Cooperative Study Group, *New England Journal of Medicine*, July 1978.

'It distributed copies . . .' *Food, Drug and Cosmetics Report*, July 1978.

'gave it their approval . . .' *Lancet*, October 1979.

p.265 'The results have been promising . . .' J. S. Meyer, *Journal of the American Geriatrics Society*, June 1989.

p.266 'More dramatically, a more recent . . .' *Neurology*, September 2002.

'medicine's search for the Holy Grail . . .' Roy Porter, *The Greatest Benefit to Mankind: A Medical History of Humanity from Antiquity to the Present* (London: HarperCollins, 1997).

p.267 'Back in the mid-1970s . . .' A. Bennett and M. Del Tacca, 'Prostaglandins in human colonic carcinoma', *Gut*, 1975.

'Many of those trials have already . . .' Gareth Morgan, *Aspirin and Cancer* (University of Wales, 2002).

p.268 'It's difficult not to be optimistic . . .' Professor Chris Paraskeva interview.

'one of the greatest finds . . .' Dr Richard Sullivan, Head of Clinical Programmes, Cancer Research UK, *BBC News*, 5 March 2003.

p.269 'In America, the FDA tried to dampen . . .' Mann and Plummer, *The Aspirin Wars*.

'once more become the top-selling analgesic . . .' *Advertising Age*, March 1988.

'It was also a time of enormous upheaval . . .' Mergers and acquisitions explained in *Pharmaceutical Review* (*Chemical and Engineering News*, December 2002).

p.271 'Around 26,000 scientific . . .' see www.ncbi.nlm.nh.gov (Pub Med: National Library of Medicine).

'Research is not restricted to the benefits of aspirin . . .' Interview with Professor Kathryn Uhrich. See also Pharmaceutical Achievers at www.chemheritage.org.

p.273 'Others have been thinking . . .' Lord, 'Polypill to fight cardiovascular disease', *British Medical Journal*, 2003.

'chewing an aspirin at the first sign of heart attack . . .' Interview with Sir Richard Peto.

'An NHS review . . .' *Sunday Times*, 13 April 2003 and interview with Gareth Morgan of the University of Wales.

'Why, then, are not more people taking it . . .' P. Elwood and M. Stillings, *Risk Factors for Vascular Disease* (University of Wales, 1999).

'It's a question of getting . . .' Sir Richard Peto interview.

p.276 'Who knows, maybe nature intended . . .' Peter Elwood.

'I have no other motive . . .' 'An account of the success of the bark of the willow in the cure of agues', *Philosophical Transactions*, Royal Society of London, 1763.

BIBLIOGRAPHY

Abramson, S., and Weissman, G., *Arthritis and Rheumatism* (January 1989)

Ackerknecht, E. H., *Medicine at the Paris Hospital, 1794–1848* (Baltimore: Johns Hopkins University Press, 1967)

Adams, A. L., *Notes of a Naturalist* (Edinburgh, 1868)

Adams, S., 'The discovery of Ibuprofen,' *Chemistry in Britain*, December 1987

Adams, S. H., 'The Great American Fraud', *Collier's Magazine*, 7 October 1905

Ambruster, H. W., *Treason's Peace: German Dyes and American Dupes* (New York: The Beecham Press, 1947)

AMIS Research Group: 'A randomised controlled trial in persons recovered from myocardial infarction', *Journal of the American Medical Association 15 February 1980*

Andrade, E. N. da C., *A Brief History of the Royal Society* (London: Royal Society, 1960)

Anon: A Faithful Narrative of the Proceedings in a Late Affair Between the Rev. Mr John Swinton, and Mr George Baker, Both of Wadham College, Oxford. to Which is Prefix'd, a Particular Account of the Proceedings Against Robert Thistlethwayte. for a Sodomitical sic Attempt Upon Mr W. French, Commoner of the Same College. (London, 1739)

Applegate, E., *Personalities and Product: A Historical Perspective on Advertising in America* (Greenwood Press, 1998)

Armstrong, H., 'Chemical Industry and Carl Duisberg', *Nature*, 22 June 1935

Aronson, S., 'The miraculous willow tree', *R.I. Med.*, June 1994

Avalos, H., *Illness and Health Care in the Ancient Near East* (Scholars Press, 1995)

Bachoffner, P., 'Two pharmacists in the Beginning of Aspirin', *Revue d'Histoire Pharmaceutique*, Paris, 1996

Bayer & Co, Letters Patent 27,088 (1898) British Patent

Bayer & Co, Patent 9123 (3 March 1900)

Bayer Company v United Drug Company (Federal Reporter, 1921)

Beaver, W., 'Analgesic development: a brief history and perspective', *Journal of Clinical Pharmacology*, April 1980

Bedford, D. B. (ed), *Oxford Encyclopaedia of Ancient Egypt: Medicine* (Oxford University Press, 2001)

Beer, J., *The Emergence of the German Dye Industry* (Illinois Studies in Social Sciences, University of Illinois Press, 1959)

Bennett, A., and Del Tacca, M., 'Prostaglandins in human colonic carcinoma', *Gut*, 1975

Berkley, G. E., *Hitler's Gift: The story of Theresienstadt* (Boston, Mass.: Branden Books, 1993)

Beveridge, W. I., *Influenza: The Last Great Plague* (Prodist, 1977)

Billings, M., *The Influenza Pandemic of 1918*, see website at www.stanford/edu/group/virus

Birch, S., *Transactions of the Royal Society of Literature* (London, 1870)

Bjorkman, E., *Our Debt to Dr Wiley* (World's Work, 1910)

Bodenbender, H. G., 'A. Eichengrün zum 80 Geburstag', *Angewandte Chemie*, 1948

Boots Pure Drug Company UK Patent Specification 971700

Borkin, J., *The Crime and Punishment of I.G. Farben* (New York: Free Press, 1978)

Born, G., 'Aggregation of blood platelets by adenosine diphosphate and its reversal', *Nature*, 1962

Boston Collaborative Drug Surveillance Group, 'Regular aspirin intake', *British Medical Journal*, March 1974

Bottero, J., *Everyday Life in Ancient Mesopotamia* (Baltimore: Johns Hopkins University Press, 2001)

Bowden, M. E., 'Tylenol: over 50 years from laboratory shelf to medicine cabinet', *Chemical Heritage*, vol. 19, 2001

Breasted, J. H., *The Edwin Smith Surgical Papyrus* (University of Chicago Press, 1930)

Bresciani-Turroni, C., *The Economics of Inflation* (London: Allen & Unwin, 1937)

Brichta, I., *The Promise and the Product: 200 years of American advertising posters* (London: Macmillan, 1979)

Brock, W., 'The Biochemical Tradition' in *Companion Encyclopaedia of the History of Medicine* (London: Routledge, 1993)

Brock, W. H., *The Fontana History of Chemistry* (London: Fontana Press, 1992)

Brodie, B., and Axelrod, J., 'The fate of acetanilide', *Journal of Pharmacology and Experimental Therapeutics*, 1949

Buchan, W., *Domestic Medicine or the Family Physician* (Edinburgh: Balfour, Auld and Smellie, 1769)

Butler, R., 'Thanks, Hippocrates, for the first miracle drug', *Geriatrics*, January 1998

Bynum, W. F., et al., *The Emergence of Modern Cardiology* (London: Wellcome Institute, Medical History 1985)

Catalogue of American Portraits (New Haven: Yale University Press, 1974)

Chaucer, Geoffrey, *The Canterbury Tales* (1387) (London: Folio Society, 1986)

Cohen, B. C., *The Ethics of Using Medical Data from Nazi Experiments* (Jewish Law: Articles 1997–2003)

Collier, H. O. J., 'The story of aspirin' *Discoveries in Pharmacology*, vol. 2: *Haemodynamics, Hormones and Inflammation*, edited by Parnham and Bruinvels (Elsevier, 1984)

Collier, H. O. J., and Shorley, P. G., *British Journal of Pharmacology*, vol. 15, 1960

Collier, H. O. J., *Scientific American*, vol. 209, 1963

Collier, R., *The Plague of the Spanish Lady* (New York: Atheneum, 1974)

Collins, S., and Lehman, J., *Excess Deaths from Influenza and Pneumonia* (New York: Public Health Monographs, 1953)

Colman Green, G., *The Archaeology and Social History of Disprin* (Hull: Reckitt, 1969)

Craven, L. L., 'Acetyl salicylic acid: possible prevention of coronary thrombosis', *Annals of Western Medicine and Surgery*, 1950

Craven, L. L., 'Experiences with aspirin in the non-specific prophylaxis of coronary thrombosis', *Mississippi Valley Medical Journal*, 1953

Craven, L. L., 'Prevention of coronary thrombosis and cerebral thrombosis', *Mississippi Valley Medical Journal*, 1956

Crawford, R., *The Spanish Flu,' Stranger than Fiction: Vignettes of San Diego History* (San Diego Historical Society, 1995)

Crosby, A. W., *Epidemic and Peace 1918: America's Forgotten Pandemic* (Cambridge University Press, 1976)

Culpeper, Nicholas, *The English Physician: or an atrologo-physical discourse of the vulgar herbs of this nation* (London: Peter Cole, 1652)

Cunningham, A., and French, R., (eds), *The Medical Enlightenment of the Eighteenth Century* (Cambridge University Press, 1990)

Dawson, W. R., and Uphill, E. P., *Who Was Who in Egyptology* (London: Egyptian Exploration Society, 1993)

Debus, A. *The Chemical Philosophy* (New York: Science History Publications, 1977)

Defoe, Daniel, *A Tour through the Whole Island of Great Britain 1724–6* (Lond. 1727)

DeKornfeld, T., Lasanga, L., and Frazier, T. M., *A Comparative Study . . .*, *Journal of the American Medical Association*, 29 December 1962

Dobson, M. J., '"Marsh Fever": the geography of malaria in England', *Journal of Historical Geography*, 1980

Doll, R., and Hill, A. B., *British Medical Journal*, 30 September 1950

Dorner, M., *Early Dye History and the Introduction of Synthetic Dyes before the 1870s* (www.smith.edu./hsc/silk/papers/dorner)

Dreser, H., *Pharmakologisches über einige Morphinderivate* (19 September 1898, Bayer Leverkusen archives)

Dreser, H., *Pharmakologisches über Aspirin-Acetylsalicylsäure* (Archiv für die Gesammte Physiologie, 1899)

Duff Gordon, Lucy, *Letters from Egypt* (London: R. Brimley Johnson, 1902)

Duisberg, C., *Meine Lebenserinnerungen* (P. Reclam. Jour., Leipzig, 1933)

Duran-Reynals, M. L., *The Fever Bark Tree: The Pageant of Quinine* (New York: Doubleday 1946)

Dyer, F., *Edison, his Life and Inventions* (New York: Harper & Bros, 1929)

Dyestuffs Committee on Ways and Means, US House of Representatives, 66th Congress. Hearings held on 18 June 1919 (US Library of Congress)

Ebell, B., *The Papyrus Ebers: The Greatest Egyptian Medical Document* (Copenhagen: Levin and Munksgaard, 1937)

Ebers, G., *Zeitschrift für Aegyptische Sprache und Alterthumskunde*, vol. 11 (Munich, 1873)

Eichengrün, A., *Pharmaceutisch-wissenschafliche Abteiling*. In: *Geschichte und Entwicklung der Farbenfabriken vorm Friedr Bayer & Co, Elberfeld, in den ersten 50 Jahren* (Munich: Meisenbach-Riffrath, 1918)

Eichengrün, A., *Dr A. Eichengrün, Aspirin, KZ Theresienstadt 1944* (Bayer Leverkusen archives)

Eichengrün, A., '50 Jahre Aspirin', *Pharmazie*, 1949

Elwood, P., and Hughes, C., 'A history of platelets, aspirin and cardiovascular disease', *Aspirin and Cardiovascular Disease*, University of Wales, 1997

Elwood, P. C., 'A randomised, controlled trial of acetyl salicylic acid in the secondary prevention of mortality from myocardial infarction', *British Medical Journal*, March 1974

Elwood, P., and Stillings, M., *Risk Factors for Vascular Disease* (University of Wales, 1999)

Elwood, P. C., and Sweetnam, P. M., 'Aspirin and secondary mortality after myocardial infarction', *Lancet*, 22 December 1979

Escales, E., *A. Eichengrün 80 Jahre* (Kunstoffe, 1947)

Ewald, P. W., *Evolution of Infectious Diseases* (Oxford University Press, 1994)

Fairley, P., *The Conquest of Pain* (London: Michael Joseph, 1978)

Farbenfabriken of Elberfeld Co. v Kuehmsted (171 Federal Reporter 1909)

Farbenfabriken Bayer AG v Sterling Drug Inc. District Court New Jersey (18 February 1960)

Farbenfabriken vormals Friedrich Bayer & Co v. Chemische Fabrik Von Heyden (Reports of Patent, Design and Trade Mark Cases, 1905, 22: 501–18)

Federal Trade Commission Decisions (*924–1935,* 921–1936, *210*–1936)

Ferguson, N., *The Pity of War* (London: Allen Lane, Penguin Press 1998)

Ferreira, S. H., Moncada, S., and Vane, J. R., 'Indomethacin and aspirin abolish prostaglandin release from spleen', *Nature* 23 June 1971

Fields, W., and Hass, W., *Aspirin Platelets and Stroke, Background for a Clinical Trial* (Warren Green, 1971)

Flechtner, H., *Carl Duisberg: vom Chemiker zum Wirtschaftsführer* (Econ. Düsseldorf, 1959)

Fleming, P., *A History of Cardiology* (Amsterdam: Rodopi, 1997).

Flexner, S., and Flexner, J. T., *William Henry Welch and the Heroic Age of American Medicine* (Baltimore: Johns Hopkins University Press, 1941)

Florey, K., *Analytical Profiles of Drug Substances*, vol. 8 (London: Academic Press, 1979)

Food and Drug Administration: *Over-the-Counter Drugs, Establishment of a Monograph for OTC Internal Analgesic, Antipyretic and Antirheumatic Products* (Federal Register, 8 July 1977)

Food and Drug Administration: *Minutes and Transcript, Cardiovascular and Renal Drugs Advisory Committee, 1 March 1983* (General Records, Rockville, Maryland)

Food and Drug Administration: *Minutes and Transcript, Cardiovascular and Renal Drugs Advisory Committee, 11 December 1984* (General Records, Rockville, Maryland)

Foster, L. G., *A Company That Cares* (New Brunswick: Johnson & Johnson, 1986)

Fox, S., *The Mirror Makers: a history of American advertising and its creators* (New York: Morrow, 1984)

Friend, D. G., *Archives of Surgery*, vol. 108, 1974

Garfield, S., *Mauve: How One Man Invented a Colour that Changed the World* (New York: Norton, 2001)

Gedetktage, S. H., 'Arthur Eichengrün 80 Jahre', *Pharmazie*, 1947

Ghalioungui, P., *The Physicians of Pharaonic Egypt* (Mainz, 1983)

Gibson, H., *Dundee Royal Infirmary 1798–1938* (Dundee: Kidd, 1948)

Gillon, S., *The Story of the 29th Division.* (London: Thomas Nelson & Sons, 1925)

Goodwin, Charles Wycliffe, *Papers*, British Library (Add, MSS 31268–980)

Grant, N., *Illustrated History of 20th Century Conflict* (London: Hamlyn, 1992)

Greene, G., *Stamboul Train* (London: Heinemann, 1932)

Grenville-Smith, R., and Barrie, A., *Aspro – how a family business grew up* (Nicholas International Ltd, 1976)

Haber, L. F., *The Chemical Industry, 1900–1930: international growth and technological change* (Oxford: Clarendon Press, 1971)

Hall, M. B., *Promoting Experimental Learning – experiment and the Royal Society 1660–1727* (Cambridge University Press, 1991)

Hamberg, M., 'Thromboxanes: a new group', *Proceedings of the National Academy of Sciences*, August 1975

Hayes, P., *Industry and Ideology. I.G. Farben in the Nazi Era* (Cambridge University Press, 1987)

Haynes, W., *American Chemical Industry – The World War I Period: 1912–1922*, vol. 2 (New York: D. Van Nostrand Company, Inc., 1945).

Hettinger, H., *A Decade of Radio Advertising* (University of Chicago Press, 1993)

Hiebert, J., *Our Policy Is People: Their Health Our Business* (New York: The Newcomers Society, 1963)

Hoehling, A. A., *The Great Epidemic* (Boston, Mass.: Little Brown, 1961)

Howell, J. D., 'Concepts of Heart-Related Diseases' in *The Cambridge World History of Human Disease* (Cambridge University Press, 1993)

'Influenza Epidemic in the British Army in France, 1918. Influenza Committee of the Advisory Board to the D.G.M.S., France', *British Medical Journal*, 1918

In the Matter of American Home Products: Federal Trade Commission (1983). *Decision*, Federal Records Centre

In the Matter of Bristol-Myers: Federal Trade Commission (1983). *Decision*, Federal Records Centre

In the Matter of Sterling Drug: Federal Trade Commission (1983). *Decision*, Federal Records Centre

Issekutz, B., *Die Geschichte der Arzneimittelforschung* (Budapest, 1971)

Jameson, E., *The Natural History of Quackery* (London: Michael Joseph, 1961)

Jay, P., *The Road to Riches* (London: Weidenfeld & Nicholson, 2000)

Jeffreys, D., *The Bureau – Inside the Modern FBI* (London: Macmillan, 1994)

Jerome, J. K., *Three Men in a Boat, 1889* (London: Bloomsbury Classics, 1997)

Jones, J. P., *The German Secret Service in America 1914–18* (Toronto: William Briggs, 1918)

King, L. S., *The Medical World of the Eighteenth Century* (University of Chicago Press, 1958)

Klein, R., 'The fever bark tree', *Natural History*, vol. 85, 1976

Kolata, G., *Flu: The Story of the Great Influenza Pandemic of 1918 and the Search for the Virus that Caused It* (New York: Farrar, Straus and Giroux, 1999)

Korthaus, W., *Pharmazeutische Geschäft in Südamerika währen des Krieges* (Bayer Leverkusen archives)

Lagnado, L. M., and Cohn Dekel, S., *Children of the Flames: Dr Josef Mengele and the untold story of the twins of Auschwitz* (London: Sidgwick and Jackson, 1991)

Landau, H., *The Enemy Within* (New York: G. P. Putnam, 1937)

Leighton, I., *The Aspirin Age* (London: Bodley Head, 1950)

Lester, D., and Greenberg, L. A., 'The metabolic fate of acetanilide', *Journal of Pharmacology and Experimental Therapeutics*, 1947

Lloyd, G. E. R. (ed.), *Hippocratic Writings* (Harmondsworth: Penguin, 1978)

Longrigg, J. N., *Greek Rational Medicine* (London: Routledge, 1991)

Luckau, A. M., *The German Delegation at the Paris Peace Conference* (New York: Columbia University Press, 1941)

Lynch, M. (ed.), *Oxford Companion to Scottish History* (Oxford University Press, 2001)

Lyons, Sir H., *The Royal Society 1660–1940* (Cambridge University Press, 1994)

Maclagan, T. J., 'Typhus statistics of the Dundee Royal Infirmary,' *Edinburgh Medical Journal*, vol. xiii, 1867–8

Maclagan, T. J., 'On enteric fever in Dundee and neighbourhood', *Edinburgh Medical Journal*, vol. xiii, 1867–8

Maclagan, T. J., 'The treatment of acute rheumatism by salicin', *Lancet*, 4 March 1876

Mann, C., and Plummer, M., *The Aspirin Wars, Money, Medicine and 100 Years of Rampant Competition* (New York: Knopf, 1991)

Mann, J., *Murder, Magic and Medicine* (Oxford University Press, 1992)

Mann, R., (entry in) *Dictionary of National Biography – Missing Persons* (Oxford University Press, 1993)

McGinnis, J. P. D., *The Impact of Epidemic Influenza in Canada* (Medicine in Canada, Historical Perspectives, 1981)

McKeown, T., *The Modern Rise of Population* (New York: Academic Press, 1976)

McTavish, J., *The German Pharmaceutical Industry 1880–1920: A Case Study of Aspirin* (Master's thesis, University of Minnesota, 1986)

McTavish, J., *Aspirin in Germany: The Pharmaceutical Industry and the Pharmaceutical Profession* (Pharmacy in History 1987)

McTavish, J., 'What's in a name? Aspirin and the American Medical Association', *Bulletin of Historical Medicine*, vol. 61, 1987

Meyer, J. S., *Journal of the American Geriatrics Society*, June 1989

Miskell, L., Whatley, C., and Harris, B., *Victorian Dundee* (Tuckwell Press, 2001)

Mitchell Palmer, A., *Aims and Purposes of The Chemical Foundation Inc and the Reasons for its Organisation. As told by A. Mitchell Palmer, United States Attorney General and Former Alien Property Custodian in his report to Congress, and by Francis P. Garvan, Alien Property Custodian, in an address to the National Cotton Manufacturers Association* (New York: De Vinne Press, 1919)

Morgan, B., *Apothecary's Venture: The Scientific Quest of the International Nicholas Organisation* (Nicholas Kiwi, 1959)

Morgan, G., *Aspirin and Cancer* (University of Wales, 2002)

Morse, H. N., *Darstellungsmethode der Aceytlamidophenole* in *Berichte der Duetchen chemischen Gesellschaft* (1878)

Mozes Kor, E., *Echoes from Auschwitz* (Candles, 1999)

Murnane, W. J., *The Princess Who Never Was: A Tale of Scholarly Agonizing, Piracy and Revenge* (Chicago: The Oriental Institute News and Notes, 1984)

Murray, D., et al., *Surgery*, August 1937

Neustadt, R., *Bacteriologie*, November 1928

Nicholson, J. S., 'Ibuprofen' in *Chronicles of Drug Discovery* (London: J. Wiley, 1982)

Nunn, J. F., *Ancient Egyptian Medicine* (London: British Museum Press, 1996)

O'Brien, J. R., 'Platelet aggregation: some results from a new method of study', *Journal of Clinical Pathology*, 1962

Order of Battle of the United States Land Forces in The World War. Zone of the Interior: Territorial Departments. Tactical Divisions Organised in 1918. Posts, Camps and Stations. (Washington, DC, Centre of Military History, US Army)

Orwell, G., *The Road to Wigan Pier* (London: Victor Gollancz, 1937)

Peto, R., et al., 'Randomised trial of prophylactic daily aspirin in British male doctors', *British Medical Journal*, 1988

Peto, R., et al., 'Antiplatelet trialists collaboration', *British Medical Journal*, 1994

Physicians' Health Study Research Group: Preliminary Report, *New England Journal of Medicine*, January 1988

Pierpoint, W. S., *Edward Stone (1702–1768) and Edmund Stone (1700–1768): confused identities resolved* (Notes Rec. Royal Society of London 51(2), 211–17, 1997)

Pierpoint, W. S., 'The natural history of salicylic acid,' *Interdisciplinary Science Reviews*, 1997

Piper, P. J., and Vane, J. R., *Nature*, 1969

Poole, J. D. F., and French, J. E., 'Thrombosis', *Journal of Atherosclerosis*, August 1961

Porter, D., and Porter, R., *Patients Progress: Doctors and Doctoring in Eighteenth Century England* (Cambridge: Polity Press, 1989)

Porter, R. *The Greatest Benefit to Mankind: A Medical History of Humanity from Antiquity to the Present* (London: HarperCollins, 1997)

Porter, R. *Blood and Guts: A Short History of Medicine*, London: Allen Lane, Penguin Press, 2002

Pyle, G. F., *The Diffusion of Influenza: Patterns and Paradigms* (New Jersey: Roman & Littlefield, 1986)

Rainsford, K. D., *Aspirin and the Salicylates* (London: Butterworth, 1994)

Ransom, A. A., Letter to *Journal of the American Medical Association*, vol. 46, 1906

Raskin, I., 'Role of salicylic acid in plants', *Annual Review of Plant Physiology, Plant Molecular Biology*, 1992

Read, A., *Kristallnacht – Unleashing the Holocaust* (London: Michael Joseph 1989)

Records of the Office of Alien Property (131), US National Archives, Washington, DC

Reimer, T. N., *Bayer & Company in the United States: German dyes. Drugs and cartels in the progressive era* (PhD. thesis, Syracuse University 1996)

Reiss, L., *Nachtrag zur innerlichen Anwendung der Salicylsäure ins besondere bei dem acuten Gelenkrheumatismus* (Berl. Klin. Wochenschr., 1876)

Reiter, P., 'From Shakespeare to Defoe: Malaria in England in the Little Ice Age', *Emerging Infectious Diseases*, vol. 6, 2000

Report of the Sanitary Condition of the Labouring Population of Scotland of 1842 (Public Records Office, Kew, London)

Rice, G., *Black November: The 1918 Influenza Epidemic in New Zealand* (London: Allen & Unwin, 1988)

Robson, R. J., *The Oxfordshire Election of 1754* (Oxford University Press, 1949)

Royal Society Journal Book, 2 June 1763

Royal Society Journal Book, 19 November 1767

Schadewaldt, H. and Alstaedter, R., *History of Pharmacological Research at Bayer* (Bayer, 1991)

Schmidt, A., *Die industrielle Chemie in ihrer Bedeutung im Weltbild und Erinnerungen an ihren Aufban* (Berlin: De Greuter, 1934)

See, G., *Bulletin Academique de Médecine*, vol. 6, Paris, 1877

Sharp, A., *The Versailles Settlement – peacemaking in Paris 1919* (London: Macmillan, 1991)

Sharp, G., *Pharmaceutical Journal*, vol. 94, 1915

Shope, R. E., 'Old, intermediate and contemporary contributions to our knowledge of pandemic influenza', *Medicine*, vol. 23, 1944

Simon, G., Deposition (Entry 199), *Records of the Office of Alien Property*, US National Archives, Washington, DC

Smith, J. B., and Willis, A. L., 'Aspirin selectively inhibits prostaglandin production in human platelets', *Nature*, 23 June 1971

Sneader, W., 'The discovery of aspirin: a reappraisal', *British Medical Journal*, 23 December 2000

Sterling, Number 10, Department of Justice Central Files, Case 60–21–56 (*Sterling Products, Inc)* RG 60 (US National Archives, Washington, DC)

Sterling-Winthrop Group Ltd v Farbenfabriken Bayer A.G. (Reports of Patent Cases, 1976)

Stewart, W. K., and Fleming, L. W., 'Perthshire pioneer of anti-inflammatory agents', *Scottish Medical Journal*, 32:141/146, 1987

Stone, E., 'An account of the success of the bark of the willow in the cure of agues', *Philosophical Transactions*, Royal Society of London, 1763

Stricker, S., *Ueber die Resultate der Behandlung der Polyartritis rheumatica mit Salicylsäure* (Berl. Klin. Wochenschr, 1876)

Taylor, A. J. P., *Origins of the Second World War* (London: Hamish Hamilton, 1961)

Taylor, B., *William Murdoch: New Lamps for Old* (London: Macmillan, 1952)

The Judgement in the Farben Trial (Bollwerk-Verlag Karl Drott, 1948)

Thorn, W. (ed.), *The Letters of Mrs Henry Adams 1865–1883* (Boston, Mass.: Little Brown, 1936)

Travis, A. S., *The Rainbow Makers: the origins of the synthetic dyestuffs industry in Western Europe* (Bethlehem: Lehigh University Press, 1983)

Trials of the War Criminals Before the Nuremberg Military Tribunals under Control Council Law 10 (Public Record Office)

Tuchman, B. W., *The Guns of August* (London: Constable, 1962)

United Drug Company v Farbenfabriken of Elberfeld (US Patent Office Cancellation No. 424)

US v Alba Pharmaceutical Company et al. (Trade Cases, 1941)

US v The Bayer Company et al. (Trade Cases, 1941)

Vane, J. R., 'The use of isolated organs for detecting active substances in the circulating blood', *British Journal of Pharmacology*, 1964

Vane, J. R., 'Inhibition of prostaglandin synthesis as a mechanism of action for aspirin-like drugs', *Nature*, 23 June 1971

Vane, J. R., and Botting, R. M., *Anti-Inflammatory drugs and their mechanism of action* (Inflammation Research, 1998)

Van Hartesveldt, F. R., *The 1918–1919 Pandemic of Influenza* (Edwin Mellen Press, 1993)

Vaughan, V. C., *A Doctor's Memories* (Indianapolis: Bobbs Merrill, 1926)

Verg, E., Plumpe, G., and Schultheis, H., *Milestones* (Bayer AG, 1988)

Wagner, B. C., *IG Auschwitz, Zwangsarbeit und Vernichtung von Haftlingen des Lagers Monowitz 1941–1945* (Munich: K.G. Saur, 2000)

Warner, T., *Landmarks in Industrial History* (London: Blackie and Son, 1909)

Weir, C., *Jesse Boot of Nottingham* (Nottingham: Boots Company, 1994)

Weiss, H. J., et al., 'The effect of the salicylates in the hemostatic properties of platelets in man', *Journal of Clinical Investigation*, 1968

Wiley, H. W., *Harvey W. Wiley – An Autobiography* (Indianapolis: Bobbs-Merrill Co., 1930)

William S. Merrell v Anacin Company US Court of Customs and Patent Appeals (1938)

Williams, C. R., 'The Place of the New York Historical Society in the growth of American interest in Egyptology', *The New York Historical Society Quarterly Bulletin*, April 1920

Wilson, J., *Signs and Wonders Upon Pharaoh: A History of American Egyptology* (University of Chicago Press, 1964)

Witthauer, K., Ther. Mh., vol. 13, 1899

Wohr F., Medical Bulletin Phil, 1902

Wohlgemut, J., Ther. Mh., vol. 13, 1899

Young, J. H., *The Toadstool Millionaires: A Social History of Patent Medicines in America before Federal Regulation* (Princeton University Press, 1961)

INDEX

A NOTE ON THE AUTHOR

Diarmuid Jeffreys is a writer, journalist and television producer who has made current affairs and documentary programmes for BBC TV, Channel 4 and others, including *Newsnight* and *The Money Programme*. He is also the author of *The Bureau: Inside the Modern FBI*. He lives with his wife and children near Lewes, East Sussex.

A NOTE ABOUT THE TYPE

The text of this book is set in Garamond 3. It is one of several versions of Garamond based on the designs of Claude Garamond. It is thought that Garamond based his font on Bembo, cut in 1495 by Francesco Griffo in collaboration with the Italian printer Aldus Manutius. Garamond types were first used in books printed in Paris around 1532. The Linotype version of Garamond from 1936 is based on the American Type Founders design by Morris Fuller Benton and Thomas Maitland Cleland. Many of the present-day versions of this type are based on the Typi Academiae of Jean Jannin cut in Sedan in 1615.

Claude Garamond was born in Paris in 1480. He learned how to cut type from his father and by the age of fifteen he was able to fashion steel punches the size of a pica with great precision. At the age of sixty, he was commissioned by King Francis I to design a Greek alphabet, for which he was given the honourable title of royal type founder. He died in 1561.